Advanced Textbooks in Control and Signal Processing

Springer
*London
Berlin
Heidelberg
New York
Barcelona
Hong Kong
Milan
Paris
Santa Clara
Singapore
Tokyo*

Other titles published in this series:

Model Predictive Control
Eduardo F. Camacho and Carlos Bordons

Introduction to Optimal Estimation
E.W. Kamen and J. Su
Publication due September 1999

Discrete-Time Signal Processing
Darrell Williamson
Publication due September 1999

K.F. Man, K.S. Tang and S. Kwong

Genetic Algorithms

Concepts and Designs

With 209 Figures

 Springer

K.F. Man, PhD
K.S. Tang, PhD
S. Kwong, PhD

City University of Hong Kong, Tat Chee Avenue, Kowloon, Hong Kong

Series Editors

Professor Michael J. Grimble, Professor of Industrial Systems and Director
Professor Michael A. Johnson, Professor of Control Systems and Deputy Director

Industrial Control Centre, Department of Electronic and Electrical Engineering, University of Strathclyde, Graham Hills Building, 50 George Street, Glasgow G1 1QE, U.K.

ISBN 1-85233-072-4 Springer-Verlag London Berlin Heidelberg

British Library Cataloguing in Publication Data
Man, K. F. (Kim F.), 1951-
  Genetic algorithms : concepts and designs
  (Advanced textbooks in control and signal processing)
  1. Genetic algorithms   2. Automatic control   3. Signal processing
  I. Title   II. Tang, K. S.   III. Kwong, S.
  629.8'95631
ISBN 1852330724

Library of Congress Cataloging-in-Publication Data
Man, K. F. (Kim F.), 1951-
  Genetic algorithms : concepts and designs / K.F. Man, K.S. Tang, and S. Kwong.
    p.   cm. -- (Advanced textbooks in control and signal processing)
  Includes bibliographical references and index.
  ISBN 1-85233-072-4 (alk. paper)
  1. Computer algorithms.   2. Genetic algorithms.   I. Tang, K. S., 1967-   .
II. Kwong, S., 1959-   .   III. Title.   IV. Series.
QA76.9.A43M36   1999   98-53817
005.1--dc21   CIP

Apart from any fair dealing for the purposes of research or private study, or criticism or review, as permitted under the Copyright, Designs and Patents Act 1988, this publication may only be reproduced, stored or transmitted, in any form or by any means, with the prior permission in writing of the publishers, or in the case of reprographic reproduction in accordance with the terms of licences issued by the Copyright Licensing Agency. Enquiries concerning reproduction outside those terms should be sent to the publishers.

© Springer-Verlag London Limited 1999
Printed in Great Britain

The use of registered names, trademarks, etc. in this publication does not imply, even in the absence of a specific statement, that such names are exempt from the relevant laws and regulations and therefore free for general use.

The publisher makes no representation, express or implied, with regard to the accuracy of the information contained in this book and cannot accept any legal responsibility or liability for any errors or omissions that may be made.

Typesetting: Camera ready by authors
Printed and bound at the Athenæum Press Ltd, Gateshead, Tyne & Wear
69/3830-543210 Printed on acid-free paper

Preface

Genetic Algorithms (GA) as a tool for a search and optimizing methodology has now reached a mature stage. It has found many useful applications in both the scientific and engineering arenas. The main reason for this success is undoubtedly due to the advances that have been made in solid-state microelectronics fabrication that have, in turn, led to the proliferation of widely available, low cost, and speedy computers.

The GA works on the Darwinian principle of natural selection for which the noted English philosopher, Herbert Spencer coined the phrase "Survival of the fittest". As a numerical optimizer, the solutions obtained by the GA are not mathematically oriented. Instead, GA possesses an intrinsic flexibility and the freedom to choose desirable optima according to design specifications. Whether the criteria of concern be nonlinear, constrained, discrete, multimodal, or NP hard, the GA is entirely equal to the challenge. In fact, because of the uniqueness of the evolutionary process and the gene structure of a chromosome, the GA processing mechanism can take the form of parallelism and multiobjective. These provide an extra dimension for solutions where other techniques may have failed completely. It is, therefore, the aim of this book to gather together relevant GA material that has already been used and demonstrated in various engineering disciplines.

With this theme in mind, the book has therefore been designed to be of interest to a wide spectrum of readers. Although the GA formulation does not rely on rigorous mathematical formulae, readers are required to acquire the fundamentals of the GA in order to fully appreciate its ability to solve problems. This assimilation process, of course, must be accompanied by a basic knowledge of computer programming techniques.

The first three chapters of this book are devoted to the mechanism of the GA in search and optimization techniques. Chapter one briefly describes the background and biological inspiration of the GA and gives simple examples. Chapter two introduces several ways in which to modify the GA formulations for application purposes. The elementary steps necessary to change the genetic operations are presented. The relationship between the objective and fitness functions to determine the quality of the GA evolutionary procedures is discussed. To further enhance the phenomena of genetic operations, an educational software game is included with this volume. An insect can be computer generated according to various defined features and specifications. A solid understanding gained from these two chapters should consolidate the reader's insight into GA utilization and allow him/her to apply this for solving problems in many engineering areas.

In Chapter three, a number of state-of-the-art techniques are introduced. These procedures are complementary to those described in Chapter two, but have the ability to solve problems that are considered complex, ill defined and sometime impossible, via the use of other gradient types of approach for search and optimization. In this chapter, the parallelism of the GA for tackling time consuming computational processes is discussed. Because the GA is not a mathematically guided scheme, it can therefore, be uniquely applied to solve multiobjective functions in a simultaneous manner. This also applies to system robustness, constrained and multimodal cost functions.

Having formulated the necessary operational procedures and techniques in previous chapters, Chapter four introduces a more advanced technique by which the GA can tackle several complicated engineering problems. In order to appreciate this method of manipulating the gene structure, the actual process of DNA formulation in the hierarchical fashion is described. The discovery of this biological finding is the direct analogy to the topology of many engineering systems. The material provided in this chapter will form an unique approach for problem solving of this nature.

Having established the GA fundamentals, Chapter five introduces the GA for solving filtering problems. Three classic cases of interest are described in this chapter and each is uniquely different in nature. Based on the development of the hierarchical GA in Chapter four, this method is introduced to tackle the H-infinity control problems discussed in Chapter six, while the same approach for the computational intelligence is given in Chapter seven.

In the context of speech utterance modeling, the GA is applied to obtain the best available training model using a dedicated parallel GA hardware architecture. Chapter eight presents a full account of this development from both software and hardware points of view.

When constraint exists within an optimization problem, this often yield sub-optimal or no solutions because mathematically guided methods may break down at discrete values. This is usually not a problem for the GA as an optimzer. Chapter nine brings out this unique GA capability to solve production planning and scheduling problems in manufacturing systems. To fall in line with the same principle, the GA is also applied for the design of communication systems. Chapter ten outlines three major designs in this area where each of the networks concerned can be minimized to the lowest order.

The works reported in this book have addressed as far as possible issues that are related to the GA. It is our belief that the fundamentals of GA material have been aptly demonstrated. We have also provided sufficient insight into various practical examples to help prevent potential pitfalls as well as highlighting the advantages of GA utilization. Readers should be able to easily assimilate the information given in this book and put the knowledge imparted to good practical use.

K F Man, K S Tang and S Kwong
City University of Hong Kong
October 1998

Acknowledgements

The authors would like to thank Ms Tina Gorman for her proof reading of the text and Ms Li Ying for implementing the GA concept contained in Chapter nine.

Table of Contents

Preface		v
1.	**Introduction, Background and Biological Inspiration**	**1**
	1.1 Biological Background	1
	1.1.1 Coding of DNA	1
	1.1.2 Flow of Genetic Information	3
	1.1.3 Recombination	5
	1.1.4 Mutation	6
	1.2 Conventional Genetic Algorithm	7
	1.3 Theory and Hypothesis	11
	1.3.1 Schema Theory	11
	1.3.2 Building Block Hypothesis	16
	1.4 A Simple Example	17
2.	**Modifications to Genetic Algorithms**	**23**
	2.1 Chromosome Representation	23
	2.2 Objective and Fitness Functions	25
	2.2.1 Linear Scaling	25
	2.2.2 Sigma Truncation	26
	2.2.3 Power Law Scaling	26
	2.2.4 Ranking	26
	2.3 Selection Methods	26
	2.4 Genetic Operations	28
	2.4.1 Crossover	28
	2.4.2 Mutation	30
	2.4.3 Operational Rates Settings	30
	2.4.4 Reordering	30
	2.5 Replacement Scheme	31
	2.6 A Game of Genetic Creatures	32
	2.7 Chromosome Representation	32
	2.8 Fitness Function	33
	2.9 Genetic Operation	38
	2.9.1 Selection Window for Functions and Parameters	38
	2.10 Demo and Run	42

3. Intrinsic Characteristics ... 45
- 3.1 Parallel Genetic Algorithm ... 45
  - 3.1.1 Global GA ... 46
  - 3.1.2 Migration GA ... 46
  - 3.1.3 Diffusion GA ... 50
- 3.2 Multiple Objective ... 51
- 3.3 Robustness ... 54
- 3.4 Multimodal ... 56
- 3.5 Constraints ... 60
  - 3.5.1 Searching Domain ... 60
  - 3.5.2 Repair Mechanism ... 60
  - 3.5.3 Penalty Scheme ... 61
  - 3.5.4 Specialized Genetic Operations ... 62

4. Hierarchical Genetic Algorithm ... 65
- 4.1 Biological Inspiration ... 66
  - 4.1.1 Regulatory Sequences and Structural Genes ... 66
  - 4.1.2 Active and Inactive Genes ... 66
- 4.2 Hierarchical Chromosome Formulation ... 67
- 4.3 Genetic Operations ... 70
- 4.4 Multiple Objective Approach ... 70
  - 4.4.1 Iterative Approach ... 71
  - 4.4.2 Group Technique ... 72
  - 4.4.3 Multiple-Objective Ranking ... 74

5. Genetic Algorithms in Filtering ... 75
- 5.1 Digital IIR Filter Design ... 75
  - 5.1.1 Chromosome Coding ... 78
  - 5.1.2 The Lowest Filter Order Criterion ... 80
- 5.2 Time Delay Estimation ... 86
  - 5.2.1 Problem Formulation ... 86
  - 5.2.2 Genetic Approach ... 87
  - 5.2.3 Results ... 92
- 5.3 Active Noise Control ... 96
  - 5.3.1 Problem Formulation ... 96
  - 5.3.2 Simple Genetic Algorithm ... 101
  - 5.3.3 Multiobjective Genetic Algorithm Approach ... 108
  - 5.3.4 Parallel Genetic Algorithm Approach ... 114
  - 5.3.5 Hardware GA Processor ... 122

6. Genetic Algorithms in H-infinity Control ... 133
- 6.1 A Mixed Optimization Design Approach ... 133
  - 6.1.1 Hierarchical Genetic Algorithm ... 137
  - 6.1.2 Application I: The Distillation Column Design ... 138
  - 6.1.3 Application II: Benchmark Problem ... 147

6.1.4 Design Comments 153

7. Hierarchical Genetic Algorithms in Computational Intelligence 155
7.1 Neural Networks 155
    7.1.1 Introduction of Neural Network 156
    7.1.2 HGA Trained Neural Network (HGANN) 158
    7.1.3 Simulation Results 163
    7.1.4 Application of HGANN on Classification 169
7.2 Fuzzy Logic.. 172
    7.2.1 Basic Formulation of Fuzzy Logic Controller 174
    7.2.2 Hierarchical Structure........................... 179
    7.2.3 Application I: Water Pump System 184
    7.2.4 Application II: Solar Plant 191

8. Genetic Algorithms in Speech Recognition Systems 199
8.1 Background of Speech Recognition Systems 199
8.2 Block Diagram of a Speech Recognition System 200
8.3 Dynamic Time Warping 203
8.4 Genetic Time Warping Algorithm (GTW) 207
    8.4.1 Encoding mechanism 208
    8.4.2 Fitness function 208
    8.4.3 Selection 209
    8.4.4 Crossover 210
    8.4.5 Mutation.. 210
    8.4.6 Genetic Time Warping with Relaxed Slope Weighting Function (GTW-RSW)............................. 211
    8.4.7 Hybrid Genetic Algorithm 212
    8.4.8 Performance Evaluation 212
8.5 Hidden Markov Model using Genetic Algorithms 216
    8.5.1 Hidden Markov Model 218
    8.5.2 Training Discrete HMMs using Genetic Algorithms ... 219
    8.5.3 Genetic Algorithm for Continuous HMM Training 225
8.6 A Multiprocessor System for Parallel Genetic Algorithms 238
    8.6.1 Implementation 241
8.7 Global GA for Parallel GA-DTW and PGA-HMM 247
    8.7.1 Experimental Results of Nonlinear Time-Normalization by the Parallel GA-DTW............................ 251
8.8 Summary... 257

9. Genetic Algorithms in Production Planning and Scheduling Problems 259
9.1 Background of Manufacturing Systems 259
9.2 ETPSP Scheme 263
    9.2.1 ETPSP Model 264

		9.2.2 Bottleneck Analysis 265

        9.2.2 Bottleneck Analysis 265
        9.2.3 Selection of Key-Processes 265
   9.3 Chromosome Configuration 266
        9.3.1 Operational Parameters for GA Cycles 266
   9.4 GA Application for ETPSP 268
        9.4.1 Case 1: Two-product ETPSP 268
        9.4.2 Case 2: Multi-product ETPSP 270
        9.4.3 Case 3: MOGA Approach 275
   9.5 Concluding Remarks 280

10. Genetic Algorithms in Communication Systems 281
   10.1 Virtual Path Design in ATM 282
        10.1.1 Problem Formulation 282
        10.1.2 Average packet delay 283
        10.1.3 Constraints 283
        10.1.4 Combination Approach............................ 284
        10.1.5 Implementation 286
        10.1.6 Results... 286
   10.2 Mesh Communication Network Design 288
        10.2.1 Design of Mesh Communication Networks 290
        10.2.2 Network Optimization using GA 291
        10.2.3 Implementation 299
        10.2.4 Results... 300
   10.3 Wireless Local Area Network Design 306
        10.3.1 Problem Formulation 306
        10.3.2 Multiobjective HGA Approach 308
        10.3.3 Implementation 310
        10.3.4 Results... 310

Appendix A 317

Appendix B 319

Appendix C 321

Appendix D 323

Appendix E 325

Appendix F 327

References 331

Index 343

1. Introduction, Background and Biological Inspiration

Our lives are essentially dominated by genes. They govern our physical features, our behaviour, our personalities, our health, and indeed our longevity. The recent greater understanding of genetics has proved to be a vital tool for genetic engineering applications in many disciplines, in addition to medicine and agriculture. It is well known that genes can be manipulated, controlled and even turned on and off in order to achieve desirable amino acid sequences of a polypeptide chain.

This significant discovery has led to the use of genetic algorithms (GA) for computational engineering. Literature concerning genetics is widely available, and therefore we are making no attempt to repeat the same information here. However, we do believe that a very brief summary of gene structure is necessary. The aim of this chapter is to outline the essential genetic phenomena which are closely associated with the formulation of GA. This includes the descriptions of the genetic operations such as crossover, mutation, selection etc. The integration of all these functions can provide a good foundation for the later developments in engineering applications. In fact, GA has proved to be a unique approach for solving various mathematical intangible problems which other gradient type of mathematical optimizers have failed to reach.

1.1 Biological Background

The fundamental unit of information in living systems is the gene. In general, a gene is defined as a portion of a chromosome that determines or affects a single character or phenotype (visible property), for example, eye colour. It comprises a segment of deoxyribonucleic acid (DNA), commonly packaged into structures called chromosomes. This genetic information is capable of producing a functional biological product which is most often a protein.

1.1.1 Coding of DNA

The basic elements of DNA are nucleotides. Due to their chemical structure, nucleotides can be classified as four different bases, Adenine (A), Guanine

(G), Cytosine (C), and Thymine (T). A and G are purines while C and T are pyrimidines. According to Watson and Crick's base pairing theory, G is paired only with C, and A is paired with T (analogue uracil* (U) in ribonucleic acid (RNA)) so that the hydrogen bonds between these pyrimidine-purine pairs are stable and sealed within the complementary strand of DNA organized in form of a double-strand helix [185], see Fig. 1.1.

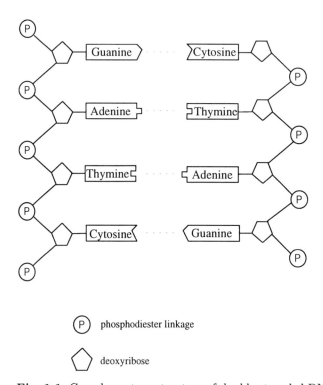

Fig. 1.1. Complementary structure of double-stranded DNA

A triplet code of nucleotide bases specifies the codon, which in turn contains a specific anticodon on transfer RNA (tRNA) and assists subsequent transmission of genetic information in the formation of a specific amino acid. Although there are 64 possible triplet codes, only 20 amino acids are interpreted by codons as tabulated in Table 1.1. It should be noticed that same amino acid may be encoded by different codons in the RNA, and, that there are three codons (UGA, UAA, and UAG) that do not correspond to any amino acid at all, but instead act as signals to stop translation (a process

---

* T is contained only in DNA and not in RNA. It will be transcribed as another nucleotide, U, in messenger RNA (mRNA).

to form polypeptide from RNA).

Table 1.1. The genetic code - from codon to amino acid

	Third element in codon			
	U	C	A	G
U U	Phenylalanine	Phenylalanine	Leucine	Leucine
U C	Serine	Serine	Serine	Serine
U A	Phenylalanine	Phenylalanine	stop	stop
U G	Cysteine	Cysteine	stop	Tryptophan
C U	Leucine	Leucine	Leucine	Leucine
C C	Proline	Proline	Proline	Proline
C A	Histidine	Histidine	Glutamine	Glutamine
C G	Arginine	Arginine	Arginine	Arginine
A U	Isolecine	Isolecine	Isolecine	Methionine
A C	Threonine	Threonine	Threonine	Threonine
A A	Asparagine	Asparagine	Lysine	Lysine
A G	Serine	Serine	Arginine	Arginine
G U	Valine	Valine	Valine	Valine
G C	Alanine	Alanine	Alanine	Alanine
G A	Aspartate	Aspartate	Glutamate	Glutamate
G G	Glycine	Glycine	Glycine	Glycine

The organizational hierarchy of DNA can be summarized as in Fig. 1.2.

1.1.2 Flow of Genetic Information

There exist three major processes in the cellular utilization of genetic information (Fig. 1.3), replication, transcription and translation.

Replication. Genetic information is preserved by DNA replication [125]. During this process, the two parent strands separate, and each serves as a template for the synthesis of a new complementary strand. Each offspring cell inherits one strand of the parental duplex; this pattern of DNA replication is described as semi-conservative.

Transcription. The first step in the communication of genetic information from the nucleus to the cytoplasm is the transcription of DNA into mRNA. During this process, the DNA duplex unwinds and one of the strands serves as a template for the synthesis of a complementary RNA strand mRNA. RNA remains single stranded and functions as the vehicle for translating nucleic acid into protein sequence.

Translation. In the process of translation, the genetic message coded in mRNA is translated on the ribosomes into a protein with a specific sequence of amino acids. Many proteins consist of multiple polypeptide chains. The

Nucleotides

Codons

Genes

DNA

Fig. 1.2. Organizational hierarchy of DNA

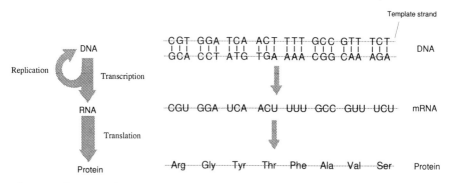

Fig. 1.3. From DNA to protein

formulation of polypeptide involves two different types of RNA namely mRNA and tRNA that play important roles in gene translation. Codons are carried by the intermediary formation of mRNA while tRNA, working as an adapter molecule, recognizes codons and inserts amino acids into their appropriate sequential positions in the polypeptide (a product of joining many amino acids). Fig. 1.4 shows the Crick's hypothesis of this translation process.

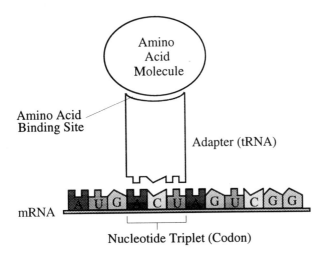

Fig. 1.4. Crick's hypothesis on tRNA

1.1.3 Recombination

Recombination is a process of the exchange of genetic information. It involves the displacement of a strand in a "host" DNA duplex by a similar strand from a "donor" duplex [174]. Pairing of the displaced host strand with the donor duplex forms a complicated Holliday structure of two duplexes lined by crossed single strands. If the appropriate single strand pairs are broken, the Holliday structure can be resolved to produce duplexes which have crossed or recombined with one another. The Holliday structure is summarized below:

1. DNA with strand break is aligned with a second homologous DNA, Fig. 1.5a;
2. Reciprocal strand switch produces a Holliday intermediate, Fig. 1.5b;
3. The crossover point moves by branch migration and strand breaks are repaired, Fig. 1.5c;
4. The Holliday intermediate can be cleaved (or resolved) in two ways, producing two possible sets of products. In Fig. 1.5d the orientation of the Holliday intermediate is changed to clarify differences in the two cleavage patterns; and

6   1. Introduction, Background and Biological Inspiration

5. The nonrecombinant and recombinant ends resulting from horizontal and vertical cleavage are shown in Figs. 1.5e and 1.5f, respectively.

Different molecular models of recombinations vary in how they postulate the structure of the host duplex, but all models are based on the ability of the invading strand to pair with its complement [210].

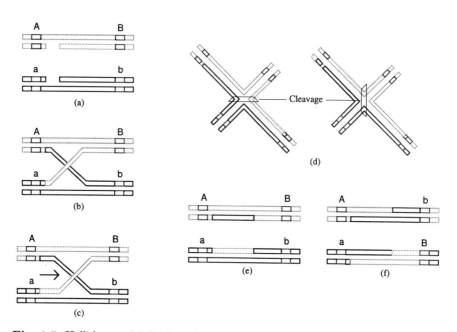

Fig. 1.5. Holliday model for homologous genetic recombination

1.1.4 Mutation

DNA is a relatively stable polymer and nucleotides generally display a very low tolerance for alterations in genetic information. Very slow, spontaneous relations such as deamination of certain bases, hydrolysis of base-sugar N-glycosidic bonds, formation of pyrimidine dimers (radiation damage), and oxidative damage are critical. An inheritable change in the phenotype; or, from the point of view of molecular biology, any change in a DNA sequence is called a mutation. In general, this operation is rare and random. The process of mutation is blind to its consequences; it throws up every possible combination of mutants, and natural selection then favours those which are better adapted to their environment. Favourable mutations that confer some advantages to the cell in which they occur are rare, being sufficient to provide the variation necessary for natural selection and thus evolution. The majority of mutations, however, are deleterious to the cell.

The most obvious cause of a mutation is an alteration in the coding sequence of the gene concerned. This is often the case with changes which affect only a single base, known as point mutation. Two kinds of point mutation exist: transition mutation and transversion mutation. In a transition mutation (Fig. 1.6b), purines are replaced by purines, and pyrimidines by pyrimidines; i.e. T–A goes to C–G or vice versa. In a transversion mutation (Fig. 1.6c), purines are replaced by purines and pyrimidines; i.e. T–A goes to A–T or G–C, C–G goes to G–C or A–T. Such mutations in coding sequences may be equally well classified by their effects. They may be neutral if there is no effect on coding properties; missence if a codon is changed to another one; or nonsense if the codon changes to a stop codon which means translation will be premature termination. In additions to point mutation, there are frameshift mutations: deletion (Fig. 1.6d), in which one or more base-pairs are lost, and insertion (Fig. 1.6e), in which one or more base-pairs are inserted into the sequence [176].

1.2 Conventional Genetic Algorithm

The basic principles of GA were first proposed by Holland [106]. Thereafter, a series of literature [47, 78, 149] and reports [18, 19, 140, 205, 221, 235] became available. GA is inspired by the mechanism of natural selection where stronger individuals are likely the winners in a competing environment. Here, GA uses a direct analogy of such natural evolution. Through the genetic evolution method, an optimal solution can be found and represented by the final winner of the genetic game.

GA presumes that the potential solution of any problem is an individual and can be represented by a set of parameters. These parameters are regarded as the genes of a chromosome and can be structured by a string of values in binary form. A positive value, generally known as a fitness value, is used to reflect the degree of "goodness" of the chromosome for the problem which would be highly related with its objective value.

Throughout a genetic evolution, the fitter chromosome has a tendency to yield good quality offspring which means a better solution to any problem. In a practical GA application, a population pool of chromosomes has to be installed and these can be randomly set initially. The size of this population varies from one problem to another although some guidelines are given in [138]. In each cycle of genetic operation, termed as an evolving process, a subsequent generation is created from the chromosomes in the current population. This can only succeed if a group of these chromosomes, generally called "parents" or a collection term "mating pool" is selected via a specific

1. Introduction, Background and Biological Inspiration

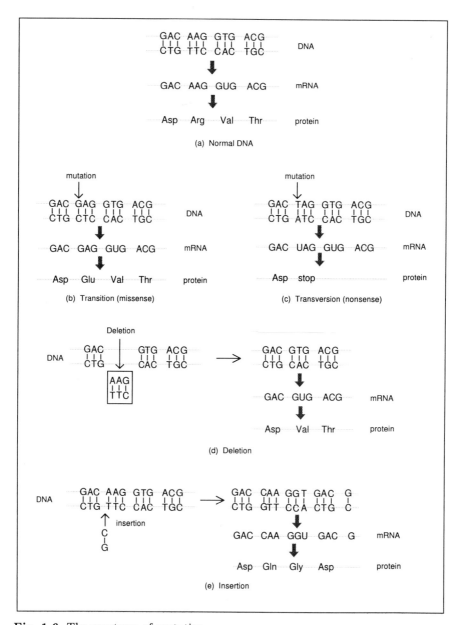

Fig. 1.6. The spectrum of mutation

selection routine. The genes of the parents are mixed and recombined for the production of offspring in the next generation. It is expected that from this process of evolution (manipulation of genes), the "better" chromosome will create a larger number of offspring, and thus has a higher chance of surviving in the subsequent generation, emulating the survival-of-the-fittest mechanism in nature.

A scheme called Roulette Wheel Selection [47] is one of the most common techniques being used for such a proportionate selection mechanism. To illustrate this further, the selection procedure is listed in Table 1.2.

Table 1.2. Roulette wheel parent selection

- Sum the fitness of all the population members; named as total fitness (F_{sum}).
- Generate a random number (n) between 0 and total fitness F_{sum}.
- Return the first population member whose fitness, added to the fitness of the preceding population members, is greater than or equal to n.

For example, in Fig. 1.7, the circumference of the Roulette wheel is F_{sum} for all five chromosomes. Chromosome 4 is the fittest chromosome and occupies the largest interval. Whereas chromosome 1 is the least fit which corresponds to a smaller interval within the Roulette wheel. To select a chromosome, a random number is generated in the interval $[0, F_{sum}]$ and the individual whose segment spans the random number is selected.

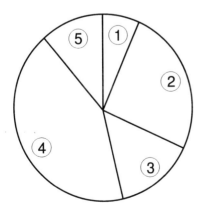

Fig. 1.7. Roulette wheel selection

The cycle of evolution is repeated until a desired termination criterion is reached. This criterion can also be set by the number of evolution cycles (computational runs), or the amount of variation of individuals between different generations, or a pre-defined value of fitness.

In order to facilitate the GA evolution cycle, two fundamental operators: Crossover and Mutation are required, although the selection routine can be termed as the other operator. To further illustrate the operational procedure, an one-point Crossover mechanism is depicted on Fig. 1.8. A crossover point is randomly set. The portions of the two chromosomes beyond this cut-off point to the right are to be exchanged to form the offspring. An operation rate (p_c) with a typical value between 0.6 and 1.0 is normally used as the probability of crossover.

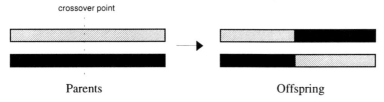

Fig. 1.8. Example of one-point crossover

However, for mutation (Fig. 1.9), this applied to each offspring individually after the crossover exercise. It alters each bit randomly with a small probability (p_m) with a typical value of less than 0.1.

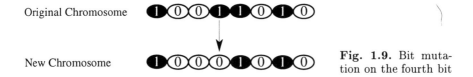

Fig. 1.9. Bit mutation on the fourth bit

The choice of p_m and p_c as the control parameters can be a complex nonlinear optimization problem to solve. Furthermore, their settings are critically dependent upon the nature of the objective function. This selection issue still remains open to suggestion although some guidelines have been introduced by [50, 86]:

– For large population size (100)
  crossover rate: 0.6
  mutation rate: 0.001

– For small population size (30)
  crossover rate: 0.9
  mutation rate: 0.01

Figs. 1.10 and 1.11 summarize the conventional GA.

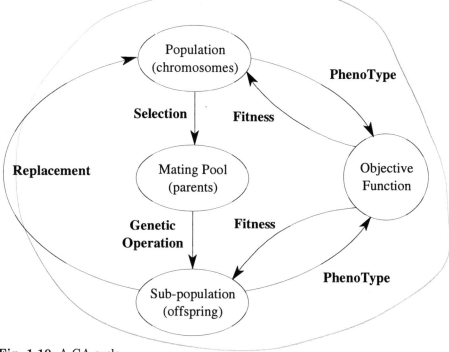

Fig. 1.10. A GA cycle

1.3 Theory and Hypothesis

In order to obtain a deeper understanding of GA, it is essential to understand why GA works. At this juncture, there are two schools of thoughts as regards its explanation: Schema Theory and Building Block Hypothesis.

1.3.1 Schema Theory

Consider a simple three-dimensional space, Fig. 1.12, and, assume that the searching space of the solution of a problem can be encoded with three bits, this can be represented as a simple cube with the string 000 at the origin. The corners in this cube are numbered by bit strings and all adjacent corners are

```
Genetic Algorithm ()
{
    // start with an initial time
    t := 0;
    // initialize a usually random population of individuals
    init_population P (t);
    // evaluate fitness of all initial individuals of population
    evaluate P (t);
    // evolution cycle
    while not terminated do
            // increase the time counter
            t := t + 1;
            // select a sub-population for offspring production
            P' := select_parents P (t);
            // recombine the "genes" of selected parents
            recombine P' (t);
            // perturb the mated population stochastically
            mutate P' (t);
            // evaluate its new fitness
            evaluate P' (t);
            // select the survivors from actual fitness
            P := survive P,P' (t);
    od
}
```

Fig. 1.11. Conventional genetic algorithm structure

1.3 Theory and Hypothesis

labelled by bit strings that differ by exactly 1-bit. If "*" represents a "don't care" or "wild card" match symbol, then the front plane of the cube can be represented by the special string 0**.

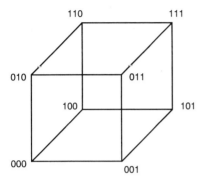

Fig. 1.12. Three-dimensional cube

Strings that contain "*" are referred to as schemata and each schema corresponds to a hyperplane in the search space. A schema represents all strings (a hyperplane or subset of the search space), which match it on all positions other than "*". It is clear that every schema matches exactly 2^r strings, where r is the number of *don't care* symbols, "*", in the schema template. Every binary encoding is a "chromosome" which corresponds to a corner in the hypercube and is a member of the $2^L - 1$ different hyperplanes, where L is the length of the binary encoding. Moreover, 3^L hyperplanes can be defined over the entire search space.

GA is a population based search. A population of sample points provides information about numerous hyperplanes. Furthermore, low order[†] hyperplanes should be sampled by numerous points in the population. A key part of the GA's intrinsic or implicit parallelism is derived from the fact that many hyperplanes are sampled when a population of strings is evaluated. Many different hyperplanes are evaluated in an implicitly parallel fashion each time a single string is evaluated, but it is the cumulative effect of evaluating a population of points that provides statistical information about any particular subset of hyperplanes.

Implicit parallelism implies that many hyperplane competitions are simultaneously solved in parallel. The theory suggests that through the process of reproduction and recombination, the schemata of competing hyperplanes increase or decrease their representation in the population according to the relative fitness of the strings that lie in those hyperplane partitions.

[†] "order" of a hyperplane refers to the number of actual bit values that appear in the schema.

Because GA operate on populations of strings, one can track the proportional representation of a single schema representing a particular hyperplane in a population. One can also indicate whether that hyperplane will increase or decrease its representation in the population over time, when fitness-based selection is combined with crossover to produce offspring from existing strings in the population. A lower bound on the change in the sampling rate of a single hyperplane from generation t to generation $t + 1$ is derived:

Effect of Selection. Since a schema represents a set of strings, we can associate a fitness value $f(S,t)$ with schema "S", and the average fitness of the schema. $f(S,t)$ is then determined by all the matched strings in the population. If *proportional selection* is used in the reproduction phase, we can estimate the number of matched strings of a schema S in the next generation.

Let $\zeta(S,t)$ be the number of strings matched by schema S at the current generation. The probability of its selection (in a single string selection) is equal to $\frac{f(S,t)}{F(t)}$, where $F(t)$ is the average fitness of the current population. The expected number of occurrences of S in the next generation is

$$\zeta(S, t+1) = \zeta(S,t)\frac{f(S,t)}{F(t)} \quad (1.1)$$

Let

$$\varepsilon = \frac{f(S,t) - F(t)}{F(t)} \quad (1.2)$$

If $\varepsilon > 0$, it means that the schema has an above average fitness and vice versa.

Substitute 1.2 into 1.1 and it shows that an "above average" schema receives an exponentially increasing number of strings in the next generations.

$$\zeta(S,t) = \zeta(S,0)(1+\varepsilon)^t \quad (1.3)$$

Effect of Crossover. During the evolution of GA, the genetic operators are disruptive to current schemata, therefore, their effects should be considered. Assuming that the length of chromosomes is L and one-point crossover is applied, in general, a crossover point is selected uniformly among $L - 1$ possible positions.

This implies that the probability of destruction of a schema S is

$$p_d(S) = \frac{\delta(S)}{L-1} \quad (1.4)$$

or the probability of a schema S survival is

$$p_s(S) = 1 - \frac{\delta(S)}{L-1} \quad (1.5)$$

1.3 Theory and Hypothesis

where δ is the *Defining Length* of the schema S defined as the distance between the outermost fixed positions.

It defines the compactness of information contained in a schema. For example, the *Defining Length* of *000* is 2, while the *Defining Length* of 1*00* is 3.

Assuming the operation rate of crossover is p_c, the probability of a schema S survival is:

$$p_s(S) = 1 - p_c \frac{\delta(S)}{L-1} \tag{1.6}$$

Note that a schema S may still survive even if a crossover site is selected between fixed positions, Eqn. 1.6 is modified as

$$p_s(S) \geq 1 - p_c \frac{\delta(S)}{L-1} \tag{1.7}$$

Effect of Mutation. If the bit mutation probability is p_m, then the probability of a single bit survival is $1 - p_m$. Denoting the *Order* of schema S by $o(S)$, the probability of a schema S surviving a mutation (i.e., sequence of one-bit mutations) is

$$p_s(S) = (1 - p_m)^{o(S)} \tag{1.8}$$

Since $p_m \ll 1$, this probability can be approximated by

$$p_s(S) \approx 1 - o(S)p_m \tag{1.9}$$

Schema Growth Equation. Combining the effect of selection, crossover, and mutation, a new form of the reproductive schema growth equation is derived

$$\zeta(S, t+1) \geq \zeta(S,t) \frac{f(S,t)}{F(t)} \left[1 - p_c \frac{\delta(S)}{L-1} - o(S)p_m\right] \tag{1.10}$$

Based on Eqn. 1.10, it can be concluded that a high average fitness value alone is not sufficient for a high growth rate. Indeed, short, low-order, above-average schemata receive exponentially increasing trials in subsequent generations of a GA.

The *Implicit Parallelism Lower Bound* derived by Holland provides that the number of schemata which are processed in a single cycle is in the order of N^3, where N is the population size. [60] derived the same result and argued that the number of schemata processed was greater than N^3 if $L \geq 64$ and $2^6 \leq N \leq 2^{20}$. This argument does not hold in general for any population size. For a particular string of length L, N must be chosen with respect

16   1. Introduction, Background and Biological Inspiration

to L to make the N^3 argument reasonable. In general, the range of values $2^6 \leq N \leq 2^{20}$ does represent a wide range of practical population sizes.

Despite this formulation, it does have its limitations that lead to the restriction of its use. Firstly, the predictions of the GA could be useless or misleading for some problems [88]. Depending on the nature of the objective function, very bad strings can be generated when good building blocks are combined. Such objective functions are referred to as GA-deceptive function [48, 77, 78]. The simplest deceptive function is the minimal deceptive problem, a two-bit function. Assuming that the string "11" represents the optimal solution, the following conditions characterize this problem:

$$
\begin{aligned}
f(1,1) &> f(0,0) \\
f(1,1) &> f(0,1) \\
f(1,1) &> f(1,0) \\
f(*,0) &> f(*,1) \quad \text{or} \quad f(0,*) > f(1,*)
\end{aligned}
\quad (1.11)
$$

The lower order schemata 0* or *0 does not contain the optimal string 11 as an instance and leads the GA away from 11. The minimal deceptive problem is a partially deceptive function, as both conditions of Eqn. 1.11 are not satisfied simultaneously. In a fully deceptive problem, all low-order schemata containing a suboptimal solution are better than other competing schemata [48]. However, [88] demonstrated that the deceptive problem was not always difficult to solve.

Secondly, the value of $f(S,t)$ in the current population may differ significantly from the value of $f(S,t)$ in the next, since schemata have interfered with each other. Thus, using the average fitness is only relevant to the first population, [89]. After this, the sampling of strings will be biased and the inexactness makes it impossible to predict computational behaviour.

1.3.2 Building Block Hypothesis

A genetic algorithm seeks near-optimal performance through the juxtaposition of short, low-order, high performance schemata, called the building block [149].

The genetic operators, we normally refer to as crossover and mutation, have the ability to generate, promote, and juxtapose (side by side) building blocks to form the optimal strings. Crossover tends to conserve the genetic information present in the strings for crossover. Thus, when the strings for crossover are similar, their capacity to generate new building blocks diminishes. Whereas mutation is not a conservative operator but is capable of generating new building blocks radically.

In addition, parent selection is an important procedure to devise. It tends to be biased towards building blocks that possess higher fitness values, and at the end ensures their representation from generation to generation.

This hypothesis suggests that the problem of coding for a GA is critical to its performance, and that such coding should satisfy the idea of short building blocks.

1.4 A Simple Example

There is no better way to show how a GA works, than to go through a real but simple example to demonstrate its effectiveness.

Problem:

Searching the global maximum point of the following objective function (see Fig. 1.13):

$$z = f(x, y)$$

where $x, y \in [-1, 1]$.

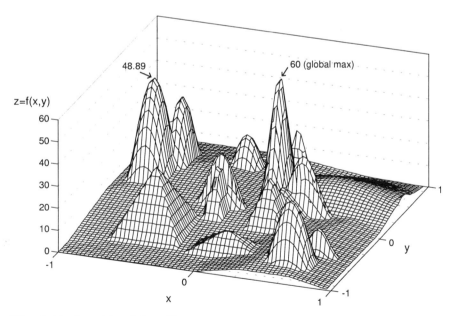

Fig. 1.13. A multimodal problem

Implementation:

The chromosome was formed by a 16-bit binary string representing x and y co-ordinates each with an eight-bit resolution. One-point crossover and bit mutation were applied with operation rates of 0.85 and 0.1, respectively. The population size was set to four for the purpose of demonstration. (In general, this number should be much larger). Here, only two offspring were generated for each evolution cycle.

This example was conducted via simulation based on MATLAB with the Genetic Toolbox [33]. Fig. 1.14 shows the typical genetic operations and the changes within the population from first to second generation.

The fitness of the best chromosome during the searching process is depicted on Fig. 1.15 and clearly shows that a GA is capable of escaping from local maxima and finding the global maximum point.

The objective values of the best chromosome in the pool against the generations are depicted in Fig. 1.16.

1.4 A Simple Example

STEP 1: Parent Selection

First Population	x	y	Objective Value z = f(x,y)
→ 0100110100101000	-0.0740	-0.6233	5.4297
→ 0101010110000101	-0.1995	0.9541	0.6696
0000010100110110	-0.9529	-0.7175	0.2562
1000101011001011	0.9070	0.1065	4.7937

STEP 2: CROSSOVER

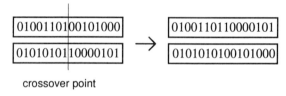

crossover point

STEP 3: MUTATION

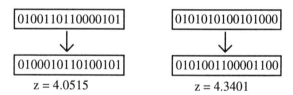

STEP 4: Reinsertion

First Population	x	y	Objective Value z = f(x,y)
0100110100101000	-0.0740	-0.6233	5.4297
→ 0100010110010101	-0.0504	0.5539	4.0515
→ 0101001100001100	-0.2309	-0.9372	4.3401
1000101011001011	0.9070	0.1065	4.7937

Fig. 1.14. Generation to generation

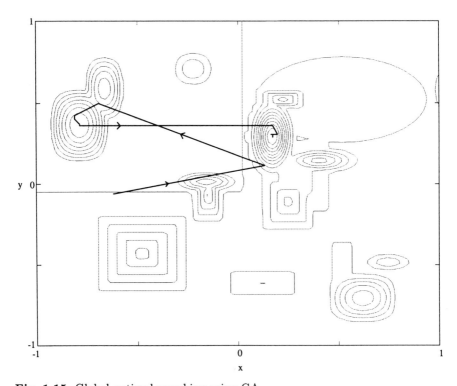

Fig. 1.15. Global optimal searching using GA

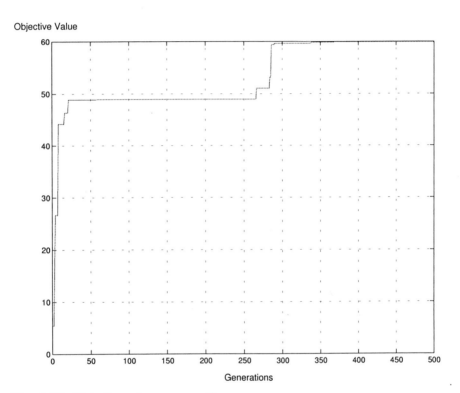

Fig. 1.16. Objective value vs generations

2. Modifications to Genetic Algorithms

The GA mechanism is neither governed by the use of differential equations nor does it behave like a continuous function. However, it possesses the unique ability to search and optimize a solution for a complex system, where other mathematical oriented techniques may have failed to compile the necessary design specifications. Due to its evolutionary characteristics, a standard GA may not be flexible enough for a practical application, and an engineering insight is always required whenever a GA is applied. This becomes more apparent where the problem to be tackled is complicated, multi-tasking and conflicting. Therefore, a means of modifying the GA structure is sought in order to meet the design requirements. There are many facets of operational modes that can be introduced. It is the main task of this chapter to outline the essential methodologies.

To further substantiate the understanding of genetic operations, this chapter also introduces an educational software game. This software has the ability to create a genetic creature of your choice, i.e. an insect. A insect will appear on the screen according to the defined specifications and features. The programme runs interactively in the Windows-95 environment. A detailed description of the software is given in the latter part of this chapter. Readers can play this game based on the layout instructions to gain further knowledge about the genetic operations.

2.1 Chromosome Representation

The problem to be tackled varies from one to the other. The coding of chromosome representation may vary according to the nature of the problem itself. In general, the bit string encoding [106] is the most classic method used by GA researchers because of its simplicity and traceability. The conventional GA operations and theory (scheme theory) are also developed on the basis of this fundamental structure. Hence, this representation is adopted in many applications. However, one minor modification can be suggested in that a Gary code may be used by the binary coding. [107] investigated the use of GA for optimizing functions of two variables based on a Gray code representation, and discovered that this works slightly better than the normal binary representation.

Recently, a direct manipulation of real-value chromosomes [114, 240] raised considerable interest. This representation was introduced especially to deal with real parameter problems. The work currently taking place [114] indicates that the floating point representation would be faster in computation and more consistent from the basis of run-to-run. At the same time, its performance can be enhanced by special operators to achieve high accuracy [149]. However, the opinion given by [79] suggested that a real-coded GA would not necessarily yield good result in some situations, despite the fact that many practical problems have been solved by using real-coded GA. So far, there is insufficient consensus to be drawn from this argument.

Another problem-oriented chromosome representation is the order-based representation which is particular useful for those problems where a particular sequence is required to search. The linear linked list is the simplest form for this representation. Normally, each node has a data field and a single link field. The link field points to the address of the successor's data field. This is a chain-like structure which is shown in Fig. 2.1.

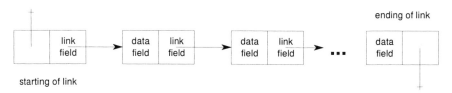

Fig. 2.1. Linear link list

This type of chromosome formulation can be found in [128] for solving the Dynamic Time Warping system and Robotic Path Planning [149] in which the sequence of the data is essential to represent a solution. In general, the length of the chromosomes in the form of link field may vary. This can be supplemented by a series of special genetic operations in order to meet the design of the GA process. A generalized structure in the form of Graph representation can be introduced. It allows loops to link the other data blocks as indicated in Fig. 2.2. A successful implementation of this structure is demonstrated by solving the graph colorings problem [47].

These order-based encoding techniques have an important advantage over literal encoding techniques in that they rule out a tremendous number of suboptimal solutions. The process avoids the problem that literal encoding encounters when illegal solutions are often generated by the crossover oper-

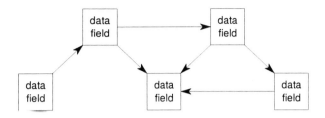

Fig. 2.2. Graphs

ations.

In some cases, an index can be used as the chromosome element instead of a real value. A typical example can be given by a look-up table format. This has proved to be a useful technique for nonlinear term selection [62].

All in all, the modification in chromosome representation comprises an endless list and is largely dependent on the nature of individual problems. A well chosen chromosome format can enhance the understanding of the problem formulation and also alleviate the burden of practical implementation.

2.2 Objective and Fitness Functions

An objective function is a measuring mechanism that is used to evaluate the status of a chromosome. This is a very important link to relate the GA and the system concerned. Since each chromosome is individually going through the same evaluating exercise, the range of this value varies from one chromosome to another. To maintain uniformity, the objective value(s) O is mapped into a fitness value(s) [78, 149] with a map Ψ where the domain of F is usually greater than zero.

$$\Psi : O \to F \tag{2.1}$$

2.2.1 Linear Scaling

The fitness value f_i of chromosome i has a linear relationship with the objective value o_i as

$$f_i = ao_i + b \tag{2.2}$$

where a and b are chosen to enforce the equality of the average objective value and the average fitness value, and cause maximum scaled fitness to be a specified multiple of the average fitness.

This method can reduce the effect of genetic drift by producing an extraordinarily good chromosome. However, it may introduce a negative fitness value which must be avoided in the GA operations [78]. Hence, the choice of a and b are dependent on the knowledge of the range of the objective values.

2.2.2 Sigma Truncation

This method avoids the negative fitness value and incorporates the problem dependent information into the scaling mechanism. The fitness value, f_i of chromosome i is calculated according to

$$f_i = o_i - (\bar{o} - c\sigma) \qquad (2.3)$$

where c is a small integer, $\bar{o}$ is the mean of the objective values, σ is the standard deviation in the population.

To prevent negative value of f, any negative result $f < 0$ is arbitrarily set to zero. Chromosomes whose fitness values are less than c (a small integer from the range 1 and 5) standard deviation from the average fitness value are not selected.

2.2.3 Power Law Scaling

The actual fitness value is taken as a specific power of the objective value, o_i

$$f_i = o_i^k \qquad (2.4)$$

where k is in general problem dependent or even varying during the run [75].

2.2.4 Ranking

There are other methods that can be used such as the Ranking scheme [10]. The fitness values do not directly relate to their corresponding objective values, but to the ranks of the objective values.

Using this approach can help the avoidance of premature convergence and speed up the search when the population approaches convergence [234]. On the other hand, it requires additional overheads in the GA computation for sorting chromosomes according to their objective values.

2.3 Selection Methods

To generate good offspring, a good parent selection mechanism is necessary. This is a process used for determining the number of trials for one particular individual used in reproduction. The chance of selecting one chromosome as

2.3 Selection Methods

a parent should be directly proportional to the number of offspring produced.

[10] presented three measures of performance of the selection algorithms, *Bias*, *Spread* and *Efficiency*.

- *Bias* defines the absolute difference between individuals in actual and expected probability for selection. Optimal zero bias is achieved when an individual's probability equals its expected number of trials.

- *Spread* is a range in the possible number of trials that an individual may achieve. If $g(i)$ is the actual number of trials due to each individual i, then the "minimum spread" is the smallest spread that theoretically permits zero bias, i.e.

$$g(i) \in \{\lfloor et(i) \rfloor, \lceil et(i) \rceil\} \qquad (2.5)$$

where $et(i)$ is the expected number of trials of individual i, $\lfloor et(i) \rfloor$ is the floor and $\lceil et(i) \rceil$ is the ceiling.
Thus, the spread of a selection method measures its consistency.

- *Efficiency* is related to the overall time complexity of the algorithms.

The selection algorithm should thus be achieving a zero bias whilst maintaining a minimum spread and not contributing to an increased time complexity of the GA.

Many selection techniques employ Roulette Wheel Mechanism (see Table 1.2). The basic roulette wheel selection method is a stochastic sampling with replacement (SSR). The segment size and selection probability remain the same throughout the selection phase and the individuals are selected according to the above procedures. SSR tends to give zero bias but potentially inclines to a spread that is unlimited.

Stochastic Sampling with Partial Replacement (SSPR) extends upon SSR by resizing a chromosome's segment if it is selected. Each time an chromosome is selected, the size of its segment is reduced by a certain factor. If the segment size becomes negative, then it is set to zero. This provides an upper bound on the spread of $\lceil et(i) \rceil$ but with a zero lower bound and a higher bias. The roulette wheel selection methods can generally be implemented with a time complexity of the order of $NlogN$ where N is the population size.

Stochastic Universal Sampling (SUS) is another single-phase sampling algorithm with minimum spread, zero bias and the time complexity in the order of N [10]. SUS uses an N equally spaced pointer, where N is the number of selections required. The population is shuffled randomly and a single random number in the range $[0, \frac{F_{sum}}{N}]$ is generated, *ptr*, where F_{sum} is the sum of the individuals' fitness values. The N individuals are then chosen by generating the N pointers spaced by 1, $[ptr, ptr+1, \ldots, ptr+N+1]$, and selecting those individuals whose fitnesses span the positions of the pointers. An individual is thus guaranteed to be selected a minimum of $\lfloor et(i) \rfloor$ times and no more than $\lceil et(i) \rceil$, thus achieving minimum spread. In addition, as individuals are selected entirely on their position in the population, SUS has zero bias.

2.4 Genetic Operations

2.4.1 Crossover

Although the one-point crossover method was inspired by biological processes, it has one major drawback in that certain combinations of schema cannot be combined in some situations [149].

For example, assume that there are two high-performance schemata:

$$\begin{aligned} S_1 &= 1\ 0\ 1\ *\ *\ *\ *\ 1 \\ S_2 &= *\ *\ *\ *\ 1\ 1\ *\ * \end{aligned}$$

There are two chromosomes in the population, I_1 and I_2, matched by S_1 and S_2, respectively:

$$\begin{aligned} I_1 &= 1\ 0\ 1\ 1\ 0\ 0\ 0\ 1 \\ I_2 &= 0\ 1\ 1\ 0\ 1\ 1\ 0\ 0 \end{aligned}$$

If only one-point crossover is performed, it is impossible to obtain the chromosome that can be matched by the following schema (S_3) as the first schema will be destroyed.

$$S_3 = 1\ 0\ 1\ *\ 1\ 1\ *\ 1$$

A multi-point crossover can be introduced to overcome this problem. As a result, the performance of generating offspring is greatly improved. One example of this operation is depicted in Fig. 2.3 where multiple crossover points are randomly selected.

Assuming that two-point crossover is performed on I_1 and I_2 as demonstrated below, the resulting offspring are shown as I_3 and I_4 in which I_3 are matched by S_3.

$$\begin{aligned} I_1 &= 1\ 0\ 1\ 1\ |\ 0\ 0\ |\ 0\ 1 \\ I_2 &= 0\ 1\ 1\ 0\ |\ 1\ 1\ |\ 0\ 0 \end{aligned}$$

$$\begin{aligned} I_3 &= 1\ 0\ 1\ 1\ \ 1\ 1\ \ 0\ 1 \\ I_4 &= 0\ 1\ 1\ 0\ \ 0\ 0\ \ 0\ 0 \end{aligned}$$

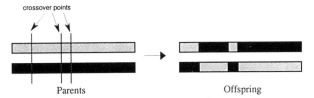

Fig. 2.3. Example of multi-point crossover

Another approach is the uniform crossover. This generates offspring from the parents, based on a randomly generated crossover mask. The operation is demonstrated in Fig. 2.4.

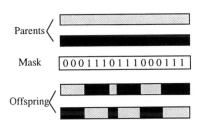

Fig. 2.4. Example of uniform crossover

The resultant offspring contain a mixture of genes from each parent. The number of effective crossing points is not fixed, but will be averaged at $L/2$ (where L is the chromosome length).

The preference for using which crossover techniques is still arguable. However, [49] concluded that a two-point crossover seemed to be an optimal number for multi-point crossover. Since then, this has been contradicted by [204] as a two-point crossover could perform poorly if the population has largely being converged because of any reduced crossover productivity. This low crossover productivity problem can be resolved by the utilization of the reduce-surrogate crossover [22].

Since the uniform crossover exchanges bits rather than segments, it can combine features regardless of their relative location. This ability may outweigh the disadvantage of destroying building block solutions and make uniform crossover superior for some problems [208]. [59] reported on several experiments for various crossover operators. A general comment was that each of these crossovers was particularly useful for some classes of problems and quite poor for others, and that the one-point crossover was considered a "loser" experimentally.

Crossover operations can be directly adopted into the chromosome with real number representation. The only difference would be if the string is composed of a series of real numbers instead of binary number.

Some other problem-based crossover techniques have been proposed. [42] designed an "analogous crossover" for the robotic trajectory generation. Therefore, the use of the crossover technique to improve the offspring production, is very much problem oriented. The basic concept in crossover is to exchange gene information between chromosomes. An effective design of crossover operation would greatly increase the convergency rate of a problem.

2.4.2 Mutation

Originally, mutation was designed only for the binary-represented chromosome. To adopt the concept of introducing variations into the chromosome, a random mutation [149] has been designed for the real number chromosome:

$$g = g + \psi(\mu, \sigma) \qquad (2.6)$$

where g is the real value gene; ψ is a random function which may be Gaussian or normally distributed; μ, σ are the mean and variance related with the random function, respectively.

2.4.3 Operational Rates Settings

The choice of an optimal probability operation rate for crossover and mutation is another controversial debate for both analytical and empirical investigations. The increase of crossover probability would cause the recombination of building blocks to rise, and at the same time, it also increases the disruption of good chromosomes. On the other hand, should the mutation probability increase, this would transform the genetic search into a random search, but would help to reintroduce the lost genetic material.

As each operator probability may vary through the generations, Davis [45] suggested linear variations in crossover and mutation probability, with a decreasing crossover rate during the run while mutation rate was increased. Syswerda [209] imposed a fixed schedule for both cases but Booker [22] utilized a dynamically variable crossover rate which was dependent upon the spread of fitness. [46, 47] modified the operator probabilities according to the success of generating good offspring. Despite all these suggested methods, the recommendation made by [50, 86] is the yardstick to follow.

2.4.4 Reordering

As stated in the building block hypothesis explained in Chap. 1, the order of genes on a chromosome is critical. The purpose of reordering is to attempt to find the gene order which has the better evolutionary potential. A technique for reordering the positions of genes in the chromosome has been suggested. The order of genes between two randomly chosen positions is inverted within the chromosome. Such a technique is known as Inversion.

For example, consider an integer represented chromosome where two inversion sites, position 3 and position 6, are chosen:

$$1 \quad 2 \quad \underbrace{3 \quad 4 \quad 5 \quad 6}_{\text{inversion region}} \quad 7 \quad 8$$

After the inversion, the order of the genes in the inversion region are reversed. Hence, we have

$$1 \quad 2 \quad 6 \quad 5 \quad 4 \quad 3 \quad 7 \quad 8$$

[45, 80, 201] combine the features of inversion and crossover into a single operator, e.g. partially matched crossover (PMX), order crossover (OX), and cycle crossover (CX).

2.5 Replacement Scheme

After generating the sub-population (offspring), several representative strategies that can be proposed for old generation replacement exist. In the case of generational replacement, the chromosomes in the current population are completely replaced by the offspring [86]. Therefore, the population with size N will generate N offspring in this strategy.

This strategy may make the best chromosome of the population fail to reproduce offspring in the next generation. So it is usually combined with an elitist strategy where one or a few of the best chromosomes are copied into the succeeding generation. The elitist strategy may increase the speed of domination of a population by a super chromosome, but on balance it appears to improve the performance.

Another modification for generational replacement is that not all of the chromosomes of the subpopulation are used for the next generation. Only a portion of the chromosomes (usually the better will win) are used to replace the chromosomes in the population.

Knowing that a larger number of offspring implies heavier computation in each generation cycle, the other scheme is to generate a small number of offspring. Usually, the worst chromosomes are replaced when new chromosomes are inserted into the population. A direct replacement of the parents by the

2. Modifications to Genetic Algorithms

corresponding offspring may also be adopted. Another way is to replace the eldest chromosomes, which stay in the population for a long time. However, this may cause the same problem as discarding the best chromosome.

2.6 A Game of Genetic Creatures

To enable a clear understanding of GA operations, particularly for newcomers to the field, an educational software package in the form of a game has been developed for a demonstration. The software is also enclosed with this book. The main theme of this software is to create a creature, i.e. a small insect, according to the creature's specifications and features. Readers can play the game by selecting various parameters of an insect provided by the programming. These can be in terms of color, shape, and size, etc. Throughout the GA evolutionary processes, which proceed according to the pre-designed selection of the basic functions such as crossover, mutation, selection and fitness etc., each stage of the evolution process will be displayed on the monitor screen until the fittest insect is formed.

This software provides an unique visual platform for learning about GA evolutionary processes while also serving as an interesting and an educational game. It is an interactive programme and runs in a Window-95 environment with the minimum requirement of an sVGA monitor sVGA (800 × 600) for display.

2.7 Chromosome Representation

An artifical insect can be coded in a chromosome with a 25-bit long binary string for representing eight different portions including the color and size. A sequence of the gene layout is shown in Fig. 2.5.

antennae	head	wing	body	feet	body color	size	head color
1	4 5	6 7	10 11	16	17 18	20 21	22 23 25

Fig. 2.5. Chromosome coding

The position of genes with their associated codings are given as listed below:

– bits 1–4 (Antennae Shape)

The first four bits in the chromosome are used to represent the antennae. A total of 16 different types of antennae can be selected and depicted in Fig. 2.6;

- bits 5–6 (Head Shape)
  The fifth and sixth bits are assigned for the representation of the shape of the head. Only four different type of heads are allowed for selection and each one is shown in Fig. 2.7;

- bits 7–10 (Wing Shape)
  A total of 16 types of wings are given with different colors attached. These are shown in Fig. 2.8.

- bits 11–16 (Body Shape)
  In order provide a wider range of insects for selection, a total of 64 types of body shapes are available. This will enlarge the selection space to avoid a rapid premature convergence. This takes up a 6-bits location and each of the body shapes is shown in Figs. 2.9–2.12;

- bits 17 (Feet)
  Only one bit is assigned for the forming of the feet. The insect has either '1' for feet or '0' for no feet. This is shown in the top portion of Fig. 2.13;

- bits 18–20 (Body Color)
  Three bits are assigned for the color of body with a total of eight different color schemes. These are shown in the middle section of Fig. 2.13;

- bits 21–22 (Size)
  The size of the insect is limited to only four types, i.e. smallest size (00), second smallest size (01), medium size (10) and biggest size(11); and

- bits 23–25 (Head Color)
  Similar to the color of the body, the head color scheme takes the the same pattern as those used for the body although the genes occupy the last two bits of the chromosome.

2.8 Fitness Function

To conceive of the intended insect using the GA evolutionary process, the feature of this particular insect (chromosome) must be specified. The programme will proceed and each of generated chromosomes will be checked according this ideally specified insect. The measure of this checking mechanism represents the fitness function. This can be a combination of the genes,

34   2. Modifications to Genetic Algorithms

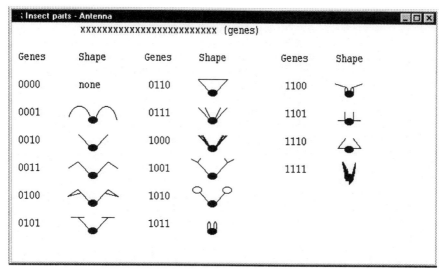

Fig. 2.6. Antennae representation

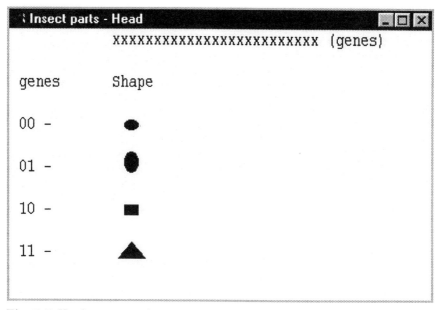

Fig. 2.7. Head representation

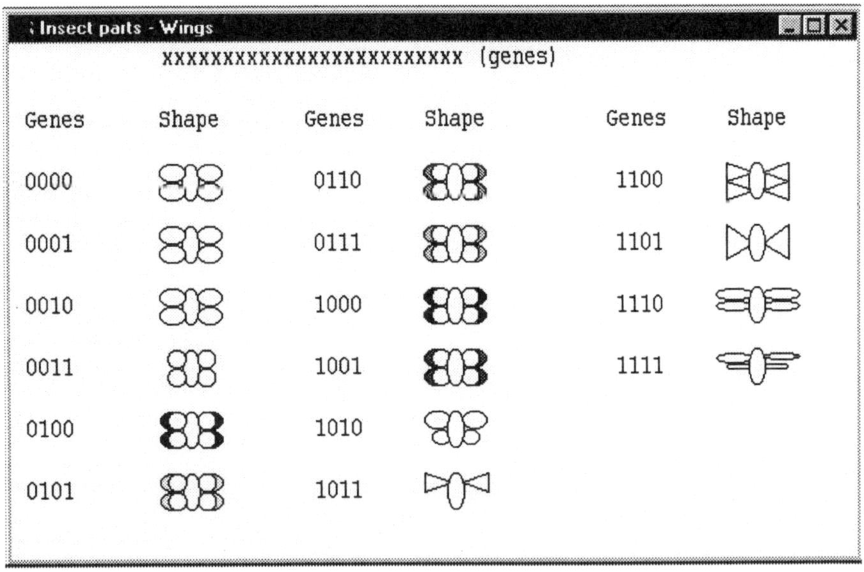

2.8 Fitness Function   35

Fig. 2.8. Wings representation

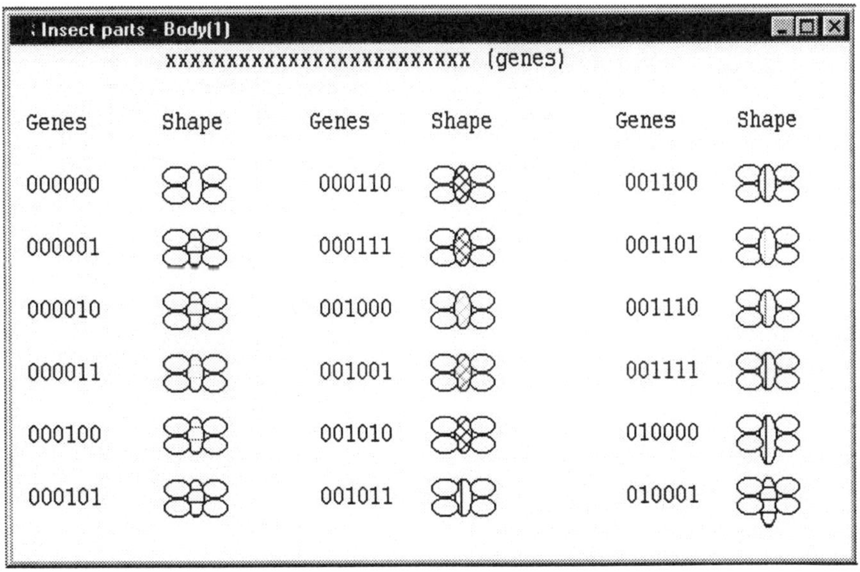

Fig. 2.9. Body representation (1)

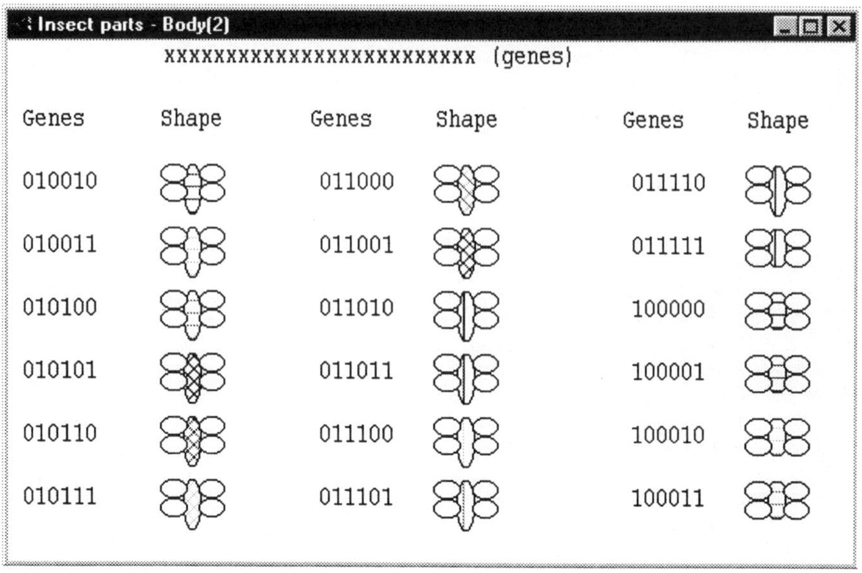

Fig. 2.10. Body representation (2)

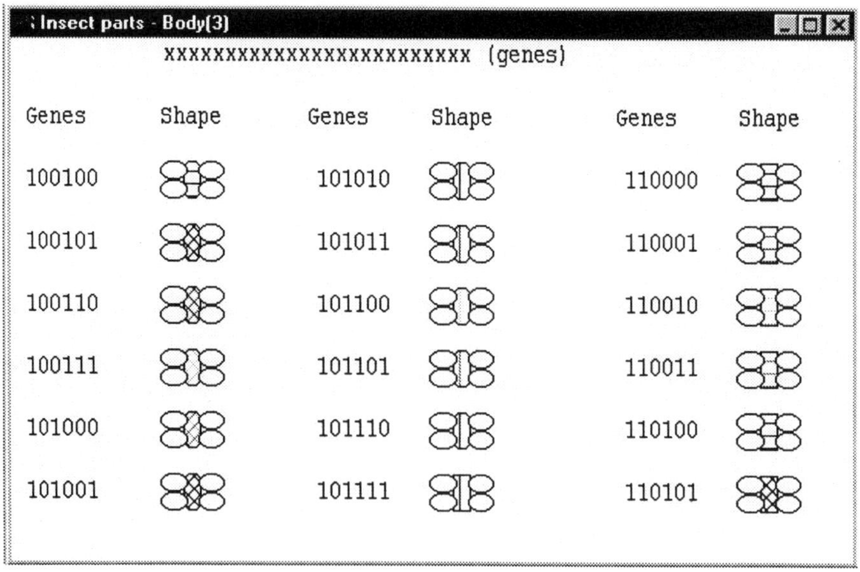

Fig. 2.11. Body representation (3)

2.8 Fitness Function   37

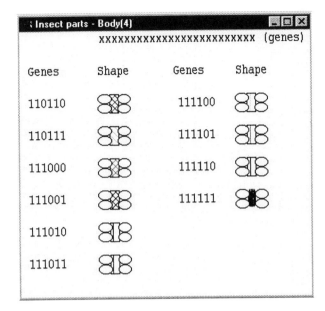

Fig. 2.12. Body representation (4)

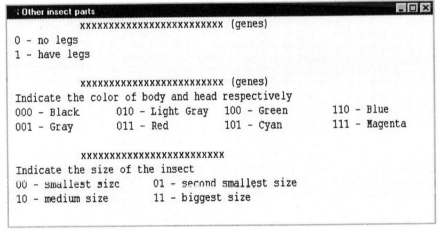

Fig. 2.13. Other appearance representation

Fig. 2.14. Insect with chromosome 0001 00 0010 110101 0 110 11 110

38   2. Modifications to Genetic Algorithms

i.e. the shape of the head, the color of the body, the shape of the wings, and size, etc. All of these can be selected in the selection window of the fitness function as depicted in Fig. 2.15.

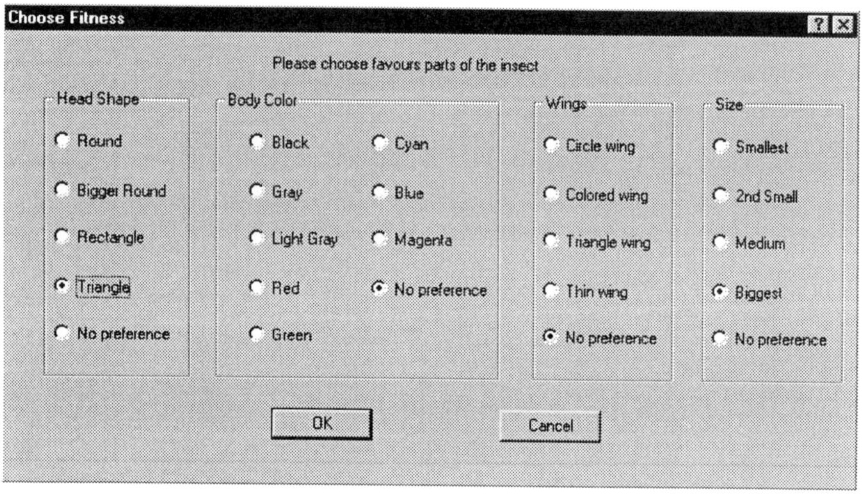

Choose Fitness				
Please choose favours parts of the insect				
Head Shape	Body Color		Wings	Size
○ Round	○ Black	○ Cyan	○ Circle wing	○ Smallest
○ Bigger Round	○ Gray	○ Blue	○ Colored wing	○ 2nd Small
○ Rectangle	○ Light Gray	○ Magenta	○ Triangle wing	○ Medium
● Triangle	○ Red	● No preference	○ Thin wing	● Biggest
○ No preference	○ Green		● No preference	○ No preference

[OK]   [Cancel]

Fig. 2.15. Selection window of fitness function

The objective functions are defined as the difference between the chosen parameters and the insect's appearance. In order to form a single fitness value, these objectives are combined by a linear weighted function.

2.9 Genetic Operation

A number of genetic operations are provided. The user can select any one of these simply by clicking the *Menu Parameter* on the toolbar. These parameters should be chosen or keyed-in before browsing the demonstration or running the program.

2.9.1 Selection Window for Functions and Parameters

Having established a fitness function for insect matching, a number of genetic functions and parameters are also required in order to carry out a proper GA evolution. All the necessary items have already been stored in the programming and can be selected by clicking the selection window. This is shown in Fig. 2.16. Based on the information provided by this figure, each of its functions is further described in the following paragraphs.

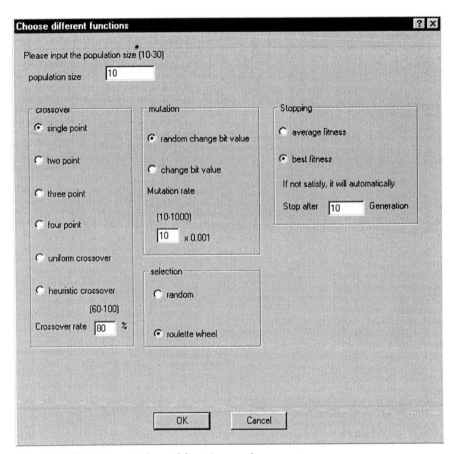

Fig. 2.16. Selection window of functions and parameters

40   2. Modifications to Genetic Algorithms

Population Size. A population pool size can be set within a range of 10 to 30 chromosomes. Due to the limitations of screen size, only a maximum of 15 insects can be displayed on the screen at one time. Therefore, when the population size exceeds 15, the first 15 insects will appear on the screen at the first incidence, followed by the second set of 15 insects after the screen has been refreshed.

Crossover. The essence of the GA operation is the method of mixing the gene structure. A total of six crossover methods are provided by this programming. The probability associated with the crossover operation is the crossover rate, which can also be selected by the user. A description of each of these methods is given below:

- *Single point crossover:*
  A single crossover point with the range [1, 24] can be randomly selected. An example of single point crossover is shown in Fig. 2.17.

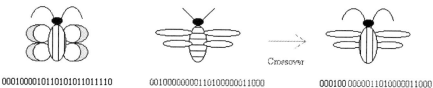

0001000010110101011011110   001000000011010000011000   000100 000001101000011000

Fig. 2.17. One point crossover

- *Multi-point crossover:*
  Rather than be restricted to the use of single point crossover, a multi-point crossover is allowed. This programming provides 2-point, 3-point and 4-point crossover which can be selected by the user. For demonstration purposes, only a chromosome (insect) with 3-point crossover is shown in Fig. 2.18. The user may use the other two crossover methods for further investigation.

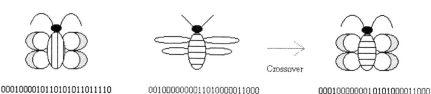

0001000010110101011011110   001000000011010000011000   0001000000010101000011000

Fig. 2.18. Three point crossover

– *Uniform crossover:*
As explained in Sect. 2.4.1, a uniform crossover mask is required during uniform crossover operations. Such a mask with the same length of the chromosome is randomly generated. In Fig. 2.19, an example of the mask "111100001000111110000000" is given to illustrate this principle.

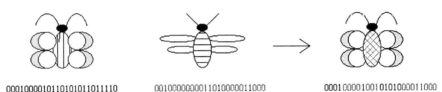

0001000010110101011011110   0010000000011010000011000   0001000010010101000011000

Fig. 2.19. Uniform crossover

– *Heuristic crossover:*
In the heuristic crossover, the binary coding is converted into integer and then computed according to the following formula to yield an offspring:

$$offspring = parent_1 \times \alpha + (1-\alpha) \times parent_2 \qquad (2.7)$$

where $\alpha \in [0,1]$. A typical result of such an operation is shown in Fig. 2.20.

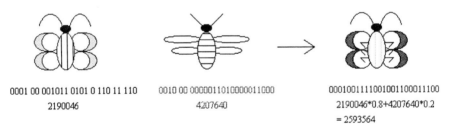

0001 00 001011 0101 0 110 11 110   0010 00 000001101000011000   000100111100100110001100
          2190046   4207640   2190046*0.8+4207640*0.2
                                                                         = 2593564

Fig. 2.20. Heuristic crossover

Mutation. Mutation is the other important genetic operation. This can prevent premature convergence from occurring. Since the binary coding is adopted in the software, then, only bit mutation is implemented. Here we have two options. The bit to be mutated is either a generated bit or is automatically flipped. The mutation rate can be set by the user. Fig. 2.21 shows the mutation operation for the change of the last bit of the chromosome.

42   2. Modifications to Genetic Algorithms

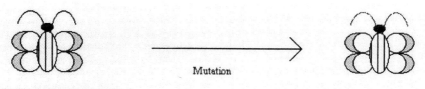

0001000010110101011011110   0001000010110101011011111

Fig. 2.21. Mutation on last bit

Selection. The parent selection has two options. These are:

1. random selection; and
2. Roulette Wheel selection

Terminational Criteria. The program is terminated either when

1. the maximum number of generations is reached. This figure is set by the user. In order to avoid long computation, the programme stops after 10 generations have provided a good indicator towards the quality of the result; or
2. the specified appearance according to the fitness function is reached by all of the insects or one of the insects in the population accounting for the choice of "average fitness" or "best fitness".

Similar information can be found at any time by clicking the *Menu Help.* the user can choose the appropriate submenu to review the meanings of various operations.

2.10 Demo and Run

In the *Menu Start*, there are two submenus:

1. *Demo* – a demonstration to show the selected crossover and mutation operations.

2. *Run* – in the submenu *Run*, there are three sub-sub-menus.
   - *Fitness* – define the preferred insects appearance as explained in Sect. 2.8
   - *Start (new pool)* – start a new GA cycle by initializing a new population pool.
   - *Run again?* – The recently terminated GA cycle may be continued by clicking this icon.

A sample run of this programme has been conducted to illustrate the purpose of this software development. The functions and parameters for this run are based on data provided in Fig. 2.16 and the subsequent figures that are associated with crossover, mutation, selection and so on. These are re-stated as follows:

1. Fitness Function
   - Head Shape   triangle
   - Body color   no preference
   - Wings   no preference
   - size   biggest
2. Functions and Parameters
   - crossover   single point
   - crossover rate   80%
   - mutation   random change bit value
   - mutation rate 10×0.001
   - selection   Roulette Wheel
   - stopping   best fitness
   - number of generations 10

A typical screen display of this run is shown in Fig. 2.22. The fittest insect obtained has a triangular head and has the biggest size. This insect matches with the criteria made in the fitness functions, which is identified and circled by the programme as shown on the screen.

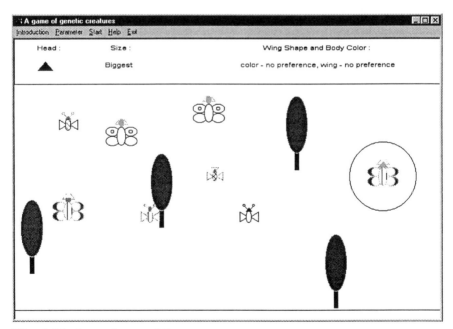

Fig. 2.22. A sample run of the program

3. Intrinsic Characteristics

Based upon the material that has been described in Chaps. 1 and 2, GA can be used to solve a number of practical engineering problems. Normally, the results obtained are quite good and are considered to be compatible to those derived from other techniques. However, a simply GA has difficulty in tackling complicated, multi-tasking and conflicting problems, and the speed of computation is generally regarded as slow. To enhance the capability of GA for practical uses, the intrinsic characteristics of GA should be further exploited and explored.

There are a number of features that have made GA become a popular tool for engineering applications. As can be understood from the previous chapters, GA is demonstrated as being a very easy-to-understand technique for reaching a solution, and that only a few, simple (sometimes no) mathematical formulations are needed. These are not the only reasons that make GA powerful optimizers, GA are also attractive for the following reasons:

- they are easy to implement in parallel architecture;
- they address multiobjective problems;
- they are capable of handling problem with constraints; and
- they can solve multimodal, non-differentiable, non-continuous or even NP-complete problems [70].

The chapter describes the full details of each item in this category, illustrating how they can be used practically in solving engineering problems.

3.1 Parallel Genetic Algorithm

Considering that the GA already possesses an intrinsic parallelism architecture, in a nutshell, there requires no extra effort to construct a parallel computational framework. Rather, the GA can be fully exploited in its parallel structure to gain the required speed for practical uses.

There are a number of GA-based parallel methods to enhance the computational speed [25, 34]. The methods of parallelization can be classified as

46   3. Intrinsic Characteristics

Global, Migration and Diffusion. These categories reflect different ways in which parallelism can be exploited in the GA as well as the nature of the population structure and recombination mechanisms used.

3.1.1 Global GA

Global GA treats the entire population as a single breeding mechanism. This can be implemented on a shared memory multiprocessor or distributed memory computer. On a shared memory multiprocessor, chromosomes are stored in the shared memory. Each processor accesses the particular assigned chromosome and returns the fitness values without any conflicts. It should be noted that there is some synchronization needed between generation to generation. It is necessary to balance the computational load among the processors using a dynamic scheduling algorithm, e.g. guided self-schedule.

On a distributed memory computer, the population can be stored in one processor to simplify the genetic operators. This is based on the farmer-worker architecture, as shown in Fig. 3.1. The farmer processor is responsible for sending chromosomes to the worker processors for the purpose of fitness evaluation. It also collects the result from them, and applies the genetic operators for the production of next generation. One disadvantage of this method is that the worker sits idly while the farmer is handling his job.

Goldberg [78] describes a modification to overcome this potential bottleneck by relaxing the requirement for strict synchronous operation. In such a case, the chromosomes are selected and inserted into the population when the worker processors complete their tasks.

Successful applications of this Global GA approach can be found in [43, 53].

3.1.2 Migration GA

This is another parallel processing mechanism for computing the GA. The migration GA (Coarse Grained Parallel GA) divides the population into a number of sub-populations, each of which is treated as a separate breeding unit under the control of a conventional GA. To encourage the proliferation of good genetic material throughout the whole population, individuals' migration between the sub-populations occurs from time to time. A pseudo code is expressed in Table 3.1.

Figs. 3.2-3.4 show three different topologies in migration. Fig. 3.2 shows the ring migration topology where individuals are transferred between directionally adjacent subpopulations. A similar strategy, known as neighbourhood

3.1 Parallel Genetic Algorithm

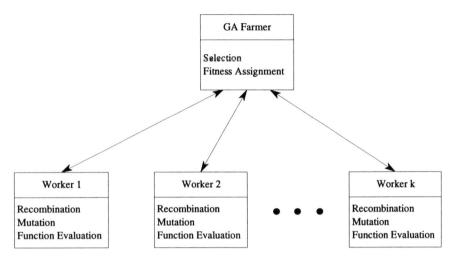

Fig. 3.1. Global GA

Table 3.1. Pseudo code of migration GA

```
- Each node (GA)

WHILE not finished
    SEQ
        Selection
        Reproduction
        Evaluation
    PAR
        Send emmigrants
        Receive immigrants
```

48   3. Intrinsic Characteristics

migration is shown in Fig. 3.3 where migration can be made between nearest neighbours bidirectionly. The unrestricted migration formulation is depicted on Fig. 3.4 where individuals may migrate from one subpopulation to the other. An appropriate selection strategy should be used to determine the migration process.

The required parameters for a successful migration depend upon:

1. *Migration Rate*
   This governs the number of individuals to be migrated, and
2. *Migration Interval*
   This affects the frequency of migrations.

The values of these parameters are intuitively chosen rather than based on some rigorous scientific analysis. In general, the occurrence of migration is usually set at a predetermined constant interval that is governed by migration intervals.

There are other approaches. [23, 155] introduce migration occurrence once the subpopulation is converged. However, the unknown quantity that determines the right time to migrate remains unsolved. This could cause the migration to occur too early. As a result, the number of correct building blocks in the migrants may be too low to influence a search on the right direction, which eventually wastes the communication resources.

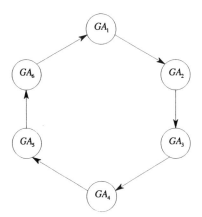

Fig. 3.2. Ring migration

The topology model of the migration GA is well suited to parallel implementation on Multiple Instruction Multiple data (MIMD) machines. The architecture of hypercubes [37, 223] and rings [83] is commonly used for this purpose. Given the range of possible population topologies and migration paths between them, efficient communication networks should thus

3.1 Parallel Genetic Algorithm   49

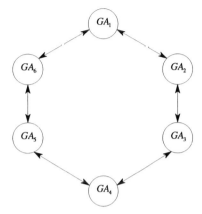

Fig. 3.3. Neighbourhood migration

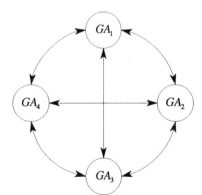

Fig. 3.4. Unrestricted migration

50   3. Intrinsic Characteristics

be possible on most parallel architecture. This applies to small multiprocessor platforms or even the clustering of networked workstations.

3.1.3 Diffusion GA

Apart from the Global and Migration techniques, Diffusion GA (Fine Grained Parallel GA) as indicated in Fig. 3.5, is the architecture can be used as a parallel GA processor. It considers the population as a single continuous structure. Each individual is assigned to a geographic location on the population surface and usually placed in a 2-D grid. This is because of the topology of the processing element in many massively parallel computers that are constructed in this form. Some other different topologies have also been studied in this area [3, 14].

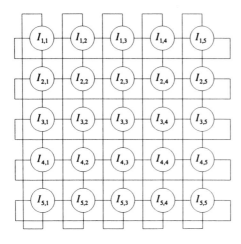

Fig. 3.5. Diffusion GA

The individuals are allowed to breed with individuals contained in a small local neighbourhood. This neighbourhood is usually chosen from immediately adjacent individuals on the population surface and is motivated by the practical communication restrictions of parallel computers. The pseudo code is listed in Table 3.2.

[141] introduced a massive parallel GA architecture with a population distributed with a 2-D mesh topology. Selection and mating were only possible with neighbouring individuals. In addition, [84, 153] introduced an asynchronous parallel GA, ASPARAGOS system. In this configuration, the GA was implemented on a connected ladder network using Transputers with one individual per processor. The practical applications of the Diffusion GA have been reported in [126] to solve a 2-D bin packing problems, and the same technique has been used to tackle a job shop scheduling problem [214].

Table 3.2. Pseudo code of diffusion GA

– Each node (I_{ij})
Initialize
WHILE not finished
SEQ
Evaluation
PAR
Send self to neighbours
Receive neighbours
Select mate
Reproduce

3.2 Multiple Objective

Without any doubt, the GA always has the distinct advantage of being able to solve multiobjective problems that other gradient type of optimizers have failed to meet. Indeed, engineering problems often exist in the class of multiple objectives. Historically, multiple objectives have been combined in an ad hoc manner so that a scalar objective function is formed for the usual linearly combined (weighted sum) functions of the multiple attributes [113, 238]. Another way is by turning the objectives into constraints. There is a lot of work being done using the weighted sums and penalty functions to turn multiobjective problems into single-attribute problems even when a GA is applied. However, a powerful method for searching the multiattribute spaces [78, 64] has been proposed to address these problems.

In this approach, the solution set of a multiobjective optimization problem consists of all those vectors such that their components cannot all be simultaneously improved. This is now known as the concept of Pareto optimality, and the solution set is called the Pareto-optimal set. The Pareto-optimal solutions are also termed as non-dominated, or non-inferior solutions, in which the definition of domination is expressed below:

Definition 3.2.1. *For an n-objective optimization problem, u is dominated by v if*

$$\forall i = 1, 2, \ldots, n, \quad f_i(u) \geq f_i(v) \quad \text{and}$$
$$\exists j = 1, 2, \ldots, n, \quad \text{such that} \quad f_i(u) > f_i(v)$$

Schaffer proposed a Vector Evaluated GA (VEGA) for finding multiple solutions to multiobjective problems [192]. This was achieved by selecting appropriate fractions of parents according to each of the objectives, separately. However, the population tends to split into species particularly strong in each

3. Intrinsic Characteristics

of the objectives if the Pareto trade-off surface is concave.

Fourman also addressed the multiple objectives in a non-aggregating manner [66]. The selection was performed by comparing pairs of individuals, each pair according to one of the objectives. The objective was randomly selected in each comparison. Similar to VEGA, this corresponds to averaging fitness across fitness components, each component being weighted by the probability of each objective being chosen to decide each tournament.

Pareto-based fitness assignment is the other method, firstly proposed by Goldberg [78]. The idea is to assign equal probability of reproduction to all non-dominated individuals in the population by using non-domination ranking and selection. He also suggested using some kind of niching to keep the GA from converging to a single point. A niching mechanism, such as sharing, would allow the GA to maintain individuals all along the trade-off surface [81]. Ritzel used a similar method, but applied deterministic crowding as the niching mechanism.

Fonseca and Fleming [63] have proposed a slightly different scheme, whereby an individual's rank corresponds to the number of individuals in the current population by which it is dominated.

$$rank(I) = 1 + p \qquad (3.1)$$

if I is dominated by other p chromosomes in the population.

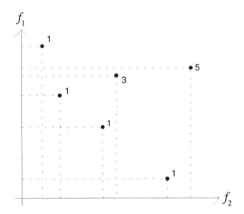

Fig. 3.6. Multiobjective ranking

Non-dominated individuals in the current population are all within the same rank, see Fig. 3.6, while dominated ones are penalized according to the population density of the corresponding region of the trade-off surface. A theory for setting the niche size is also presented in [63].

This rank-based fitness can also include the goal information in which ranking is based on the preferable rank. Consider two chromosomes I_a and I_b with $F(I_a) = [f_{a,1}, f_{a,2}, \ldots, f_{a,m}]$ and $F(I_b) = [f_{b,1}, f_{b,2}, \ldots, f_{b,m}]$, and the goal vector $V = (v_1, v_2, \ldots, v_m)$ where v_i is the goal for the design objective f_i.

Case 1: $F(I_a)$ meet none of goals
$F(I_a)$ is preferable to $F(I_b)$ $\Leftrightarrow$
$F(I_a)$ is partially less than $F(I_b)$, $F(I_a)$ $p<$ $F(I_b)$, i.e.

$$\forall i = 1, 2, \ldots, m, \quad f_{a,i} \leq f_{b,i} \wedge \exists j = 1, 2, \ldots, m, \quad f_{a,j} < f_{b,j}$$

Case 2: $F(I_a)$ meet all of goals
$F(I_a)$ is preferable to $F(I_b)$ $\Leftrightarrow$

$$F(I_a) \; p< \; F(I_b) \vee \sim (F(I_b) \leq V)$$

Case 3: $F(I_a)$ partially meet the design goals
Without loss of generality, let

$$\exists k = 1, 2, \ldots, m-1, \quad \forall i = 1, 2, \ldots, k,$$
$$\forall j = (k+1), (k+2), \ldots, m,$$
$$(f_{a,i} > v_i) \wedge (f_{a,j} \leq v_j)$$

$F(I_a)$ is preferable to $F(I_b)$ $\Leftrightarrow$

$$[(f_{a,(1,2,\ldots,k)} \; p< \; f_{b,(1,2,\ldots,k)}) \vee (f_{a,(1,2,\ldots,k)} = f_{b,(1,2,\ldots,k)})] \wedge$$
$$[(f_{a,(k+1,k+2,\ldots,m)} \; p< \; f_{b,(k+1,k+2,\ldots,m)}) \vee$$
$$\sim (f_{b,(k+1,k+2,\ldots,m)} \leq v_{(k+1,k+2,\ldots,m)})]$$

Tournament Selection based on Pareto dominance has also been proposed in [109]. In addition to the two individuals competing in each tournament, a number of other individuals in the population were used to determine whether the competitors were dominated or not. In the case where both competitors were either dominated or non-dominated, the result of the tournament was decided through sharing.

The advantage of Pareto-ranking is that it is blind to the convexity or the non-convexity of the trade-off surface. Although the domination of certain species may still occur if certain regions of the trade-off are simply easier to find than others, Pareto-ranking can eliminate sensitivity to the possible non-convexity of the trade-off surface. Moreover, it rewards good performance in any objective dimension regardless of others. Solutions which exhibit good performance in many, if not all, objective dimensions are more likely to be produced by recombination [135].

Pareto-based ranking correctly assigns all non-dominated individuals the same fitness, but that, on its own, does not guarantee that the Pareto set can be uniformly sampled. When presented with multiple equivalent optima, finite populations tend to converge to only one of these, due to stochastic errors in the selection process. This phenomenon, known as genetic drift, has been observed in natural as well as in artificial evolution, and can also occur in Pareto-based evolutionary optimization.

The additional use of fitness sharing [81] was proposed by Goldberg to prevent genetic drift and to promote the sampling of the whole Pareto set by the population. Fonseca and Fleming [63] implemented fitness sharing in the objective domain and provided a theory for estimating the necessary niche sizes, based on the properties of the Pareto set. Horn and Nafpliotis [109] also arrived at a form of fitness sharing in the objective domain. In addition, they suggested the use of a metric combining of both the objective and the decision variable domains, leading to what was called nested sharing.

The viability of mating is another aspect which becomes relevant as the population distributes itself around multiple regions of optimality. Different regions of the trade-off surface may generally have very different genetic representations, which, to ensure viability, requires mating to happen only locally [78]. So far, mating restriction has been implemented based on the distance between individuals in the objective domain [63, 92].

3.3 Robustness

There are many instances where it is necessary to make the characteristics of the system variables adaptive to dynamic signal behaviour, and ensure that they are capable of sustaining the environmental disturbance. These often require an adaptive algorithm to optimize time-dependent optima which might be difficult to obtain by a conventional GA. When a simple GA is being used, the diversity of the population is quickly eliminated as it seeks out a global optimum. Should the environment change, it is often unable to redirect its search to a different part of the space due to the bias of the chromosomes. To improve the convergency of the standard GA for changing environments, two basic strategies have been developed.

The first strategy expands the memory of the GA in order to build up a repertoire of ready responses to environmental conditions. A typical example in this group is Triallelic representation [82]. Triallelic representation consists of a diploid chromosome and a third allelic structure for deciding dominance.

The random immigrants mechanism [87] and the triggered hypermutation mechanism [35, 36] are grouped as another type of strategy. This approach increases diversity in the population to compensate for the changes encountered in the environment. The random immigrants mechanism is used to replace a fraction of a conventional GA's population, as determined by the replacement rate, with randomly generated chromosomes. It works well in environments where there are occasional, large changes in the location of the optimum.

An adaptive mutation-based mechanism, known as the triggered hypermutation mechanism, has been developed to adapt to the environmental change. The mechanism temporarily increases the mutation rate to a high value (hypermutation rate) whenever the best time-average performance of the population deteriorates.

A simulation has been conducted for illustrating the response of the GA to environmental changes. The task was to locate the global maximum peak, numerically set to 60, for the landscape depicted in Fig. 1.13. It had two-variable functions and each variable was represented in 16 bits. In other words, each population member was 32 bits long. The other parameter settings of the GA are tabulated in Table 3.3.

Table 3.3. Parameter settings of conventional GA

Representation	16-bit per variable (total 32 bit)
Population size	100
Generation gap	0.8
Fitness assignment	ranking
Selection	roulette wheel selection
Crossover	one-point crossover
Crossover rate	0.6
Mutation	bit mutation
Mutation rate	0.001

Environmental changes were introduced to testify as to the robustness of the conventional GA and the mechanisms, Random Immigrant Mechanism and Hypermutation. The changes were as follows:

1. linear translation of all of the hills in the first 50 generations. The hill location was increased by one step in both dimensions after five

generations. Each dimension's rate of change was specified independently, so that one dimension might increase while another was decreased; and
2. relocation of the maximum hill randomly every 20 generations in the period of 50-150 generations, while keeping the remainder of the landscape fixed.

Table 3.4 summarizes the simulation results.

Table 3.4. Simulation results

	Parameter setting	Result
Conventional GA	Table 3.3	Fig. 3.7
Random Immigrant	Replacement rate = 0.3	Fig. 3.8
Hypermutation	Hypermutation rate = 0.1	Fig. 3.9

In Fig. 3.7, it can be observed that the conventional GA is unable to relocate the global maximum. As explained before, this is due to the lack of population diversity. It can be revealed by comparing the average fitness and the best fitness in which these two values are approximately the same.

The Random Immigrant Mechanism and Hypermutation performed well in the experiment. However, it is worth noting that Hypermutation is triggered only when a change in the environment decreases the value of the best of the current population. In some situations, for example, if a new optimum, of say 70, occurred in another area with all the others remaining unaltered, there would be no triggers and the Hypermutation would be unlikely to detect the change.

Moreover, in a noisy condition, it is difficult to determine a change of environment. Statistical process control is hence proposed [216] to monitor the best performance of the population so that the GA-based optimization system adapts to the continuous, time-dependent nonstationary environment. The actual implementation can be referred to in the case study described in Chap. 5.1.

3.4 Multimodal

The other attribute of the GA is its capability for solving multimodal problems. Three factors [78] contribute to its ability to locate a global maximum:

– searching from a population of solutions, not a single one

3.4 Multimodal 57

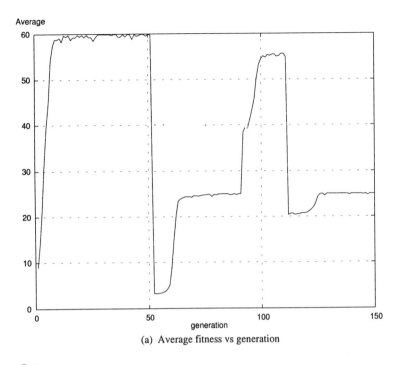

(a) Average fitness vs generation

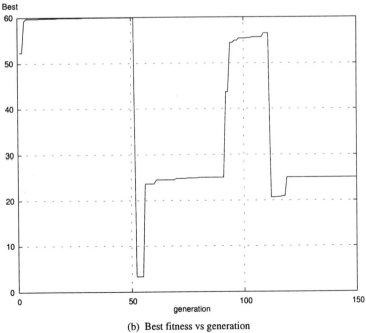

(b) Best fitness vs generation

Fig. 3.7. Conventional GA

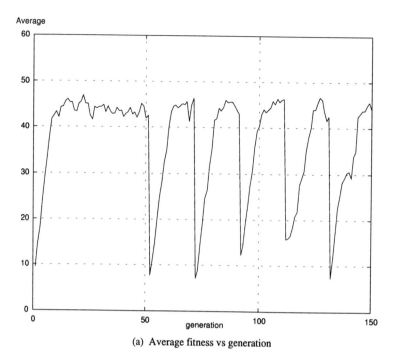

(a) Average fitness vs generation

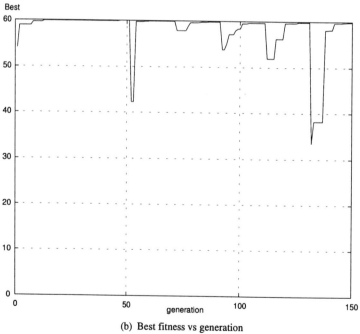

(b) Best fitness vs generation

Fig. 3.8. Random immigrant mechanism

3.4 Multimodal 59

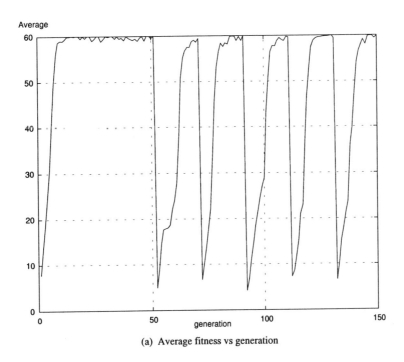

(a) Average fitness vs generation

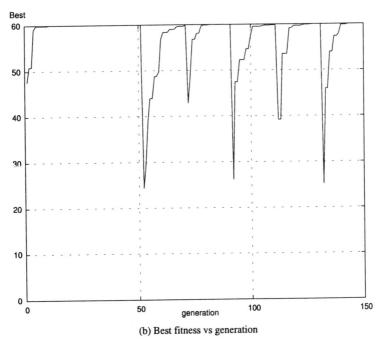

(b) Best fitness vs generation

Fig. 3.9. Hypermutation

60   3. Intrinsic Characteristics

- using fitness information, and not derivatives or other auxiliary information; and
- using randomized operators, not deterministic ones.

However, there is no guarantee that the global optimal point will be obtained by using GA although there is the tendency for this to occur. The possibility of success is reduced if there is a loss of population diversity. As the GA has a tendency to seek a sub-optimal point, the population may converge towards this value which leads to premature convergence. The global optimal solution is only obtained via the exploration of mutation in the genetic operations. Such a phenomenon is known as a Genetic Drift [22] and this situation occurs easily with a small size of population.

A number of techniques have been proposed to limit the effect of genetic drift and maintain the population diversity. These include Preselection [137], Crowding [49, 78] and Fitness Sharing [65].

3.5 Constraints

In the process of optimization, the problem of constraint is often encountered. This obstacle is not always handled properly by the conventional, but mathematically governed optimization techniques. By contrast, constraints present no problems to the GA and various methods can be used in this area.

3.5.1 Searching Domain

It is possible to embed the condition of constraints in the system by confining the searching space of a chromosome. This approach guarantees that all chromosomes are valid and that the constraint will not be violated. A typical example is to limit the searching domain of the coefficients of a digital lattice filter design in the range of $+1$ to -1 whose pole locations will be confined within the unit circle for stability. This method of solving the constraint problem requires no additional computing power, and all chromosomes created are regarded as potential solutions to the problem.

3.5.2 Repair Mechanism

The repair mechanism is a direct analogy to the DNA repair systems in the cell. In DNA, there are numerous repair systems for cell survival and the diverse sources of DNA damage. These systems may cure mismatching problem, abnormal bases, etc. This process can be emulated by the GA to solve the constraint problem. If any condition of the constraint is violated by a chromosome, the chromosome will be *"corrected"* so that it becomes valid.

This can be achieved by modifying some genes randomly within the valid solution space, or backtracking toward its parents' genetic material [121]. However, this method is rather computationally intensive. It behaves in the same way as DNA, where repairing is extraordinarily inefficient in an energetic sense. Nevertheless, it is a worthy direction to approach since the fitness evaluation is normally the most time consuming process. The repair mechanism can ensure that each chromosome undertaking the fitness evaluation is a potential solution for the constrained problem.

3.5.3 Penalty Scheme

Another approach for handling constraints is to set up a penalty scheme for invalid chromosomes such that they become low performers. The constrained problem is then transformed to an unconstrained condition by associating the penalty with all the constraint violations. This can be done by including a penalty to adjust the optimized objective function.

Consider Eqn. 3.2 as the original objective function to be optimized,

$$f(x_1, x_2, \ldots, x_n) \tag{3.2}$$

To comprise a penalty scheme within the objective function, Eqn. 3.2 becomes

$$f(x_1, x_2, \ldots, x_n) + \delta \sum_{i=1}^{m} \varphi_i \tag{3.3}$$

where m is the total number of constraints; δ is a penalty coefficient which is negative for maximization and positive for minimization problems; and φ_i is a penalty related to the i-th constraint (i=1,2,...,m).

The penalty scheme has two distinct characteristics:

- some vital information may be thrown away; and
- a small violation of a constraint may qualify if it produces a large payoff in other areas.

However, an appropriate penalty function is not easy to come by and affects the efficiency of the genetic search [179]. Moreover, computing time is wasted in evaluating the invalid chromosome, especially when the problem is one in which the constraints are likely to be violated.

3.5.4 Specialized Genetic Operations

Michalewicz introduced another direction for handling the constrained numerical problem based on his software package GENOCOP (Genetic Algorithm for Numerical Optimization with Linear Constraints) [149]. The floating point representation was used and the crossover and mutation operations were specifically designed so that the newly generated offspring were included in the solution space.

Mutation. For a convex space S, every point $s_0 \in S$ and any line p such that $s_0 \in p$, p intersects the boundaries of S at precisely two points, denoted by $l_p^{s_0}$ and $u_p^{s_0}$.

3 different mutation operations are designed based on this characteristic.

1. For uniform mutation, the mutating gene v_k which is the k-th component of chromosome s_t is a random value from the range $[l_{(k)}^{s_t}, u_{(k)}^{s_t}]$ where $l_{(k)}^{s_t}$ and $u_{(k)}^{s_t}$ are the k-th components of the vectors $l_p^{s_t}$ and $u_p^{s_t}$, respectively and $l_{(k)}^{s_t} \leq u_{(k)}^{s_t}$.
2. For boundary mutation, the mutating gene v_k is either $l_{(k)}^{s_t}$ or $u_{(k)}^{s_t}$ with equal probability.
3. For non-uniform mutation, the mutating gene v_k is modified as

$$v_k = \begin{cases} v_k + \Delta(t, u_{(k)}^{s_t} - v_k) & \text{if a random digit is 0} \\ v_k - \Delta(t, v_k - l_{(k)}^{s_t}) & \text{if a random digit is 1} \end{cases}$$

Crossover. [149] designed three crossovers based on another characteristic of the convex space:

For any two points s_1 and s_2 in the solution space S, the linear combination $a \cdot s_1 + (1-a) \cdot s_2$, where $a \in [0,1]$, is a point in S.

Consider two chromosomes $s_v^t = [v_1, \ldots, v_m]$ and $s_w^t = [w_1, \ldots, w_m]$ crossing after k-th position,

1. For single crossover, the resulting offspring are

$$s_v^{t+1} = [v_1, \ldots, v_k, w_{k+1} \cdot a + v_{k+1} \cdot (1-a),$$
$$\ldots, w_m \cdot a + v_m \cdot (1-a)]$$
$$s_w^{t+1} = [w_1, \ldots, w_k, v_{k+1} \cdot a + w_{k+1} \cdot (1-a),$$
$$\ldots, v_m \cdot a + w_m \cdot (1-a)]$$

2. For single arithmetical crossover, the resulting offspring are

$$s_v^{t+1} = [v_1, \ldots, v_{k-1}, w_k \cdot a + v_k \cdot (1-a), v_{k+1}, \ldots, v_m]$$
$$s_w^{t+1} = [w_1, \ldots, w_{k-1}, v_k \cdot a + w_k \cdot (1-a), w_{k+1}, \ldots, w_m]$$

3. For whole arithmetical crossover, the resulting offspring are

$$s_v^{t+1} = a \cdot s_w^t + (1-a) \cdot s_v^t$$
$$s_w^{t+1} = a \cdot s_v^t + (1-a) \cdot s_w^t$$

This is an effective method for handling constraints for numerical problems, but there is a limitation in solving non-numerical constraints such as the topological constraints found in networking [164].

4. Hierarchical Genetic Algorithm

Thus far, the essence of the GA in both theoretical and practical domains has been well demonstrated. The concept of applying a GA to solve engineering problems is feasible and sound. However, despite the distinct advantages of a GA for solving complicated, constrained and multiobjective functions where other techniques may have failed, the full power of the GA in engineering application is yet to be exploited and explored.

To bring out the best use of the GA, we should explore further the study of genetic characteristics so that we can fully understand that the GA is not merely a unique technique for solving engineering problems, but that it also fulfils its potential for tackling scientific deadlocks that, in the past, were considered impossible to solve. In this endeavour, we have chosen as our target an examination of the biological chromosome structure. It is acknowledged in biological and medical communities that the genetic structure of a chromosome is formed by a number of gene variations, that are arranged in a hierarchical manner. Some genes dominate other genes and there are active and inactive genes. Such a phenomenon is a direct analogy to the topology of many engineering systems that have yet to be determined by an adequate methodology.

In light of this issue, a reprise of biological genetics was carried out. A method has been proposed to emulate the formulation of a biological DNA structure so that a precise hierarchical genetic structure can be formed for engineering purposes. This translation from a biological insight enables the development of an engineering format that falls in-line with the usual GA operational modes of action. The beauty of this concept is that the basic genetic computations are maintained. Hence, this specific genetic arrangement proves to be an immensely rich methodology for system modelling, and its potential uses in solving topological scientific and engineering problems should become a force to be reckoned with in future systems design.

This chapter outlines this philosophical insight of genetic hierarchy into an engineering practice in which all the necessary techniques and functionalities are described. The achievements that can be derived from this method for solving typical engineering system topology designs are given in the following chapters.

4.1 Biological Inspiration

4.1.1 Regulatory Sequences and Structural Genes

The biological background of the DNA structure has already been given in Chap. 1. An end product, generally known as a polypeptide, is produced by the DNA only when the DNA structure is experiencing biological and chemical processes. The genes of a complete chromosome may be combined in a specific manner, in which there are some active and inactive genes. It has become a recognized fact that there are 4,000 genes in a typical bacterial genome, or an estimated 100,000 genes in a human genome. However, only a fraction of these genes in either case will be converted into an amino acid (polypeptide) at any given time and the criteria for a given gene product may change with time. It is therefore crucial to be able to regulate the gene expression in order to develop the required cellular metabolism, as well as to orchestrate and maintain the structural and functional differences between the existing cells.

From such a genetic process, the genes can thus be classified into two different types:

– regulatory sequences (RSs) and
– structural genes (SGs)

The SGs are coded for polypeptides or RNAs, while the RSs serve as the leaders that denote the beginning and ending of SGs, or participate in turning on or off the transcription of SGs, or function as initiation points for replication or recombination. One of the RSs found in DNA is called the promoter, and this activates or inactivates SGs due to the initialization of transcription. This initialization is governed by the Trans-acting Factor (TAF) [54, 118, 142] acting upon this sequence in the DNA. The transcription can only be taken place if a particular TAF is bound on the promoter. A polypeptide is then produced via a translation process in which the genetic message is coded in mRNA with a specific sequence of amino acids. Therefore, a hierarchical structure is obtained within a DNA formation that is depicted in Fig. 4.1.

4.1.2 Active and Inactive Genes

One of the most surprising discoveries in the founding of molecular biology was that active and inactive genes exist in the SGs. The active genes are

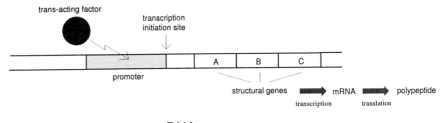

Fig. 4.1. Trans-acting factor bound on promoter for the initiation of transcription

separated into non-contiguous pieces along the parental DNA. The pieces that code mRNA are referred to as exons (active genes) and the non-coding pieces are referred as introns (inactive genes). During transcription, there is a process of splicing, Fig. 4.2 so that the final messenger RNA, which contains the exons only, are formed.

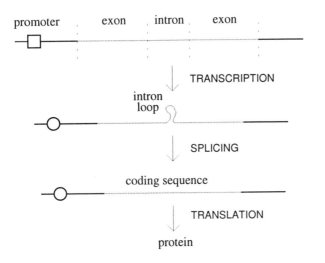

Fig. 4.2. Splicing

4.2 Hierarchical Chromosome Formulation

It is the existence of the promoter together with the commanding signals of TAF that ignites the innovated introduction of hierarchical formulation of the chromosome for GA. This chromosome can be regarded as the DNA that has already been described, but consists of the parametric genes (analogy to structural genes in DNA) and the control genes (analogy to regulatory sequences in DNA). This architecture is very similar to that shown in Fig. 4.1.

4. Hierarchical Genetic Algorithm

To generalize this architecture, a multiple level of control genes are introduced in a hierarchical fashion as illustrated in Fig. 4.3. In this case, the activation of the parametric gene is governed by the value of the first-level control gene, which is governed by the second-level control gene, and so on.

To indicate the activation of the control gene, an integer "1" is assigned for each control gene that is being ignited where "0" is for turning off. When "1" is signalled, the associated parameter genes due to that particular active control gene are activated in the lower level structure. It should be noticed that the inactive genes always exist within the chromosome even when "0" appears. This hierarchical architecture implies that the chromosome contains more information than that of the conventional GA structure. Hence, it is called Hierarchical Genetic Algorithm (HGA).

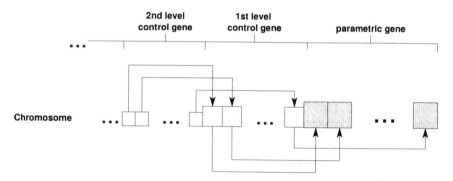

Fig. 4.3. Hierarchical chromosome structure

The use of the HGA is particularly important for the structure or topology as well as the parametric optimization. Unlike the set-up of the conventional GA optimization, where the chromosome and the phenotype structure are assumed to be fixed or pre-defined, HGA operates without these constraints. To illustrate this concept further, the following example is used to demonstrate the functionality of the HGA for engineering applications.

Example 4.2.1. Consider a chromosome formed with 6-bit control genes and 6-integer parametric genes as indicated by Fig. 4.4:

The length of X_A and X_B are 4 and 2, respectively, which means that the phenotype in different lengths is available within the same chromosome formulation. Then, the HGA will search over all possible lengths including the parameters so that the final objective requirement is met. Moreover, the hierarchical levels can be increased within the chromosome to formulate a multilayer chromosome. This is shown in Fig. 4.5 where a three-level gene structure is represented.

4.2 Hierarchical Chromosome Formulation

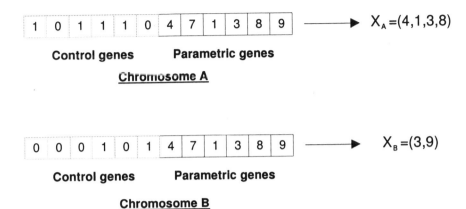

Fig. 4.4. Example of HGA chromosome representation

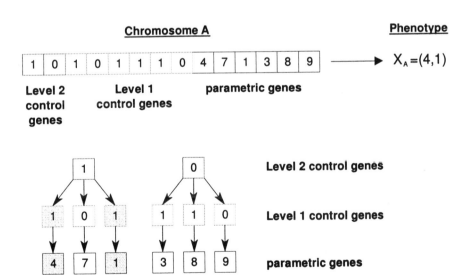

Fig. 4.5. An example of a 3-level chromosome

4.3 Genetic Operations

Since the chromosome structure of HGA is fixed, and this is true even for different parameter lengths, there is no extra effort required for reconfiguring the usual genetic operations. Therefore, the standard methods of mutation and crossover may apply independently to each level of genes or even for the whole chromosome if this is homogenous. However, the genetic operations that affect the high-level genes can result in changes within the active genes which eventually leads to a multiple change in the lower level genes. This is the precise reason why the HGA is not only able to obtain a good set of system parameters, but can also reach a minimized system topology.

4.4 Multiple Objective Approach

The basic multiple objective approaches have already been described in Chap. 3. In general, the same approach applies to the HGA. Since its main purpose is to determine the topology of the system, an extra objective function should be included for optimization. Therefore, besides the objective functions $\left(F_i = \begin{bmatrix} f_1 & f_2 & \cdots & f_i \end{bmatrix}^T\right)$ that have been defined by the problem settings as introduced before, another objective (f_{i+1}) is installed for topology optimization.

Based on the specific HGA chromosome structure, the topology information can be acquired from the control genes. Hence, by including the topology information as objective, the problem is now formulated as a multiobjective optimization problem:

$$F = \begin{bmatrix} F_i \\ f_{i+1} \end{bmatrix} \tag{4.1}$$

The GA operation is applied to minimize this objective vector, i.e.

$$\min_{x \in \Phi} F(x) \tag{4.2}$$

where Φ is the searching domain of the chromosome x.

In general, the complexity of the topology can be qualified with integer number, "N". It is assumed that a smaller value of N means a lower order structure which is more desirable.

In order to combine the topological and parametric optimization in a simultaneous manner, let us consider a candidate x_i which is depicted in Fig. 4.6. The candidate x_i has $F_i(x_j) = M_j = \begin{bmatrix} m_1 & m_2 & \cdots & m_i \end{bmatrix} \neq \underline{0}$ and

$f_{i+1}(x_j) = n_j$, and is not a solution for the problem since $\exists k$ s.t. $m_k > 0$. The solution set for the problem is represented by $\{x : F_i(x) = 0 \text{ and } N_1 \leq f_2(x) \leq N_2\}$. The best solution is denoted by x_{opt} where and $f_{i+1}(x_{opt}) = N_1$. Another solution is to have $f_{i+1}(x_1) = n_1 > N_1$, but, this runs the risk of having a higher order of complexity for the topology.

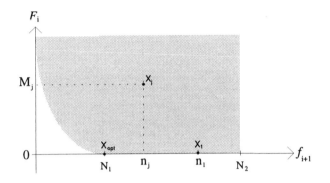

Fig. 4.6. Solution set for topology and parmeteric optimization problem

In order to obtain an optimal solution x_{opt}, both (F_i) and $(f_i + 1)$ should therefore be considered simultaneously. Various methods are developed for searching this optimal result.

4.4.1 Iterative Approach

The main difficult in topology optimization is that the degree of complexity of the topology is not known. Therefore, in order to reach an optimal solution, an iterative approach is proposed. The procedure is listed as follows:

1. Let N_2 be the maximum allowable topology for searching. The HGA is applied and terminated when a solution x_1 with $(F_i = 0)$ is obtained. (Fig. 4.7a).
2. Assuming that $f_{i+1}(x_1) = N_3$, the searching domain for the complexity of the topology is reduced from N_2 to $N_3 - 1$. The HGA is then applied again until another solution with $(F_i = 0)$ is obtained. (Fig. 4.7b)
3. Repeat Step 2 by reducing the searching domain of the topology complexity and eventually the optimal point x_{opt} with $f_{i+1}(x_{opt}) = N_1$ will be obtained. (Fig. 4.7c)
4. Another iteration with the complexity of the topology bounded by $N_1 - 1$ is carried out and, of course, no solution may be found. This process can be terminated by setting a maximum number of generations for the HGA. If no solution is found after the generation exceeds this maximum number, the solution obtained in step 3 would be considered as the optimal solution for the problem with lowest complexity. (Fig. 4.7d)

4. Hierarchical Genetic Algorithm

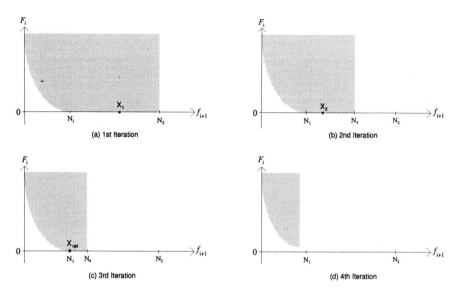

Fig. 4.7. Iterative approach for obtaining optimal solution with lowest complexity

4.4.2 Group Technique

Within a population P, chromosomes can be divided into several sub-groups, G_i, containing the same topology chromosomes (see Fig. 4.8). Each sub-group will contain less than or equal to λ chromosomes, where λ is a pre-defined value, hence,

$$size(G_i) \leq \lambda \quad \forall i = 1, 2, \ldots, M \tag{4.3}$$

The total population size at k-th generation and the maximum population size are expressed in Eqns. 4.4 and 4.5, respectively.

$$P^{(k)} = G_1^{(k)} \cup G_2^{(k)} \cdots \cup G_M^{(k)} \tag{4.4}$$

$$P_{max} = \lambda M \tag{4.5}$$

In order to manage these sub-groups, a modified insertion strategy is developed. The top level description of the insertion strategy for new chromosome z is expressed in Table 4.1.

This grouping method ensures that the chromosomes compete only with those of the same complexity in topological size (f_{i+1}). Therefore, the topology information will not be lost even if the other objective values (F_i) are poor. The selection pressure will still rely on the objective values (F_i) to the problem.

4.4 Multiple Objective Approach

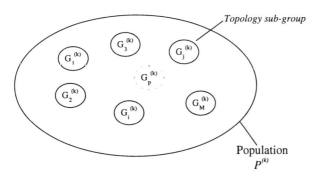

Fig. 4.8. Sub-groups in population P

Table 4.1. Insertion strategy

At generation $(k+1)$
Step 1:
If $\left\{ G^{(k)}_{f_{i+1}(z)} = \emptyset \quad \text{or} \quad size\left[G^{(k)}_{f_{i+1}(z)} \right] < \lambda \right\}$ then
$\quad \left\{ G^{(k+1)}_{f_{i+1}(z)} = G^{(k)}_{f_{i+1}(z)} \cup \{z\} < \lambda \right\}$
else
$\quad$ goto step 2
Step 2:
If $\left\{ F_i(z) < F_m = \max\left\{ F_i(z_j), \; \forall z_j \in G^{(k)}_{f_{i+1}(z)} \right\} \right\}$ then
$\quad \left\{ G^{(k+1)}_{f_{i+1}(z)} = \left\{ z_j : F_i(z_j) < F_m, \; z_j \in G^{(k)}_{f_{i+1}(z)} \right\} \cup \{z\} \right\}$
else
$\quad$ goto step 3
Step 3:
$\quad$ Exit

74   4. Hierarchical Genetic Algorithm

4.4.3 Multiple-Objective Ranking

Instead of using the fitness value for selection, a ranking method involving multiple objective information [63] can be adopted and hence the information of the topology is included.

Chromosome I is then ranked by

$$rank(I) = 1 + p \tag{4.6}$$

if I is dominated by other p chromosomes in the population. The fitness assignment procedure is listed as below:

1. Sort population according to preferable rank;
2. Assign fitness to individual by interpolating from the highest rank to the lowest, and a linear function is used

$$f(I) = f_{min} + (f_{max} - f_{min}) \frac{rank(I) - 1}{N_{ind} - 1} \tag{4.7}$$

   where f_{min} and f_{max} are the lower and upper limits of fitness, respectively; and N_{ind} is the population size.
3. Average the fitness of individual in the same rank, so that all of them will be selected at the same rate.

5. Genetic Algorithms in Filtering

In filtering design, we often encounter the dilemma of choosing a minimum order of a model for the representation of a signal so that a minimum use of the computational power is ensured. In the case of real time processing, this requirement is necessary and obvious. Achieving this goal is by no means easy. This can be very involved in circumstances where the performance criteria are discrete, complex, multiobjective and very often, nonlinear.

This chapter outlines the main features of GA applications in these areas and demonstrates the GA capability for solving problems of this nature. For the case of infinite-impulse-response (IIR) filter design, the employment of HGA together with a multiobjective approach is used to find a minimum order of a IIR filter while the involving performance requirements are fully met.

As for discrete time delay estimation problems, GA is applied to estimate the discrete time delay parameter of a nonlinear time series function. This method is then extended to the active noise control application, whereby in this particular design, not only is each noise reduction scheme clearly brought out, but hardware specific architecture is also developed for the computation of time consuming GA operations.

5.1 Digital IIR Filter Design

The traditional approach to the design of discrete-time IIR filters involves the transformation of an analogue filter into a digital filter at a given set of prescribed specifications. In general, a bilinear transformation is adopted. Butterworth (BWTH), Chebyshev Type 1 (CHBY1), Chebyshev Type 2 (CHBY2) and Elliptic (ELTC) function based on approximation methods [39, 91, 129, 228] are the most commonly used techniques for implementing the frequency-selective analogue IIR filters. The order of filter is normally determined by the magnitude of the frequency response. The obtained order is considered to be minimal but applicable only to a particular type of filter. This technique for designing the digital filter can be found from the use of

76   5. Genetic Algorithms in Filtering

MATLAB toolbox [146].

To design the other types of frequency-selective filters such as highpass (HP), bandpass (BP), and bandstop (BS) filters, an algebraic transformation is applied so that the design of HP, BP or BS filters can be derived upon the prototype design of a lowpass (LP) filter [39]. From this transformation, the order of BP and BS filters are found to be twice as much as the LP filter.

However, very often, when we come to the design of a digital IIR filter, an optimal filtering performance is much preferred. This generally consists of the following constraints which are strictly imposed upon the overall design criteria, i.e.

- the determination of lowest filter order;
- the filter must be stable (the poles must lie inside the unit circle); and
- must meet the prescribed tolerance settings.

The first item is not easily determined by any formal method, despite the great amount of effort that has been spent in this area. As for the following two points, these are the constraints which the final filter has to satisfy in order to meet the design performance. These constraints may pose great difficulty for the purpose of optimization in order to meet the design criteria.

Having realized the advantages of HGA formulation, realizing the transformation of the IIR design criteria into HGA presents no problems. Unlike the classic methods, there is no restriction on the type of filter for the design. Each individual filter, whether it is LP, HP, BP or BS, can be independently optimized until the lowest order is reached. This is only possible because of the intrinsic property of the HGA for solving these demanding functions in a simultaneous fashion and, one that involves no an extra cost and effort. To this end, the use of HGA for IIR filter design is summarized as follows:

- the filter can be constructed in any form, such as cascade, parallel, or lattice;
- the LP, HP, BP and BS filters can be independently designed;
- no mapping between analogue-to-digital is required;
- multiobjective functions can be simultaneously solved; and
- the obtained model is of the lowest order.

To illustrate this scheme in detail, Fig. 5.1 depicts the four types of frequency responses that have to be satisfied with various tolerance settings (δ_1, δ_2). The objective of the design is to ensure that these frequency responses are confined within the prescribed frequencies (ω_p, ω_s).

To proceed with the design optimally, the objective functions (f_1, f_2) should be specified in accordance to the specifications of the digital filters.

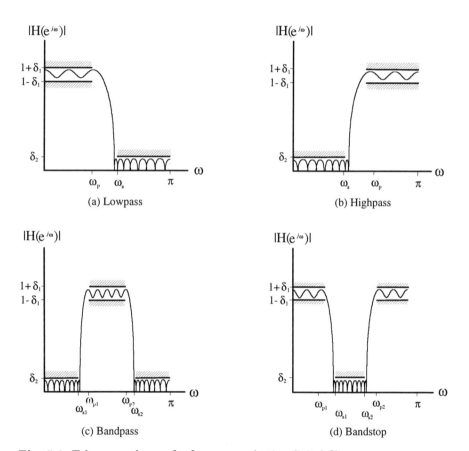

Fig. 5.1. Tolerance schemes for frequency-selective digital filter

78   5. Genetic Algorithms in Filtering

These variables are tabulated in Table 5.1 for each filter structure where $\Delta H_\omega^{(p)}$ and $\Delta H_\omega^{(s)}$ are computed as below:

$$\Delta H_\omega^{(p)} = \begin{cases} |H(e^{j\omega})| - (1 + \delta_1) & \text{if } |H(e^{j\omega})| > (1 + \delta_1) \\ (1 - \delta_1) - |H(e^{j\omega})| & \text{if } |H(e^{j\omega})| < (1 - \delta_1) \end{cases} \quad (5.1)$$

and

$$\Delta H_\omega^{(s)} = \left|H(e^{(j\omega)})\right| - \delta_2 \quad \text{if } \left|H(e^{(j\omega)})\right| > \delta_2 \quad (5.2)$$

Table 5.1. Objective functions

Filter Type	Objective Functions	
	$f_1 = \sum \Delta H_\omega^{(p)}$	$f_2 = \sum \Delta H_\omega^{(s)}$
LP	$\forall \omega : 0 \leq \omega \leq \omega_p$	$\forall \omega : \omega_s \leq \omega \leq \pi$
HP	$\forall \omega : \omega_p \leq \omega \leq \pi$	$\forall \omega : 0 \leq \omega \leq \omega_s$
BP	$\forall \omega : \omega_{p1} \leq \omega \leq \omega_{p2}$	$\forall \omega : 0 \leq \omega \leq \omega_{s1}, \quad \omega_{s2} \leq \omega \leq \pi$
BS	$\forall \omega : 0 \leq \omega \leq \omega_{p1}, \quad \omega_{p2} \leq \omega \leq \pi$	$\forall \omega : \omega_{s1} \leq \omega \leq \omega_{s2}$

5.1.1 Chromosome Coding

The essence of the HGA is its ability to code the system parameters in a hierarchical structure. Despite the format of chromosome organization, this may have to change as, from one system to the other, the arrangement of gene structures can be set in a multilayer fashion. As for the case in IIR filter design, the chromosome representation of a typical pulse transfer function $H(z)$ as indicated in Fig. 5.2 is

$$H(z) = K \frac{(z + b_1)(z + b_2)(z^2 + b_{11}z + b_{12})}{(z + a_1)(z + a_2)(z^2 + a_{11}z + a_{12})} \quad (5.3)$$

where K is a gain; and $a_i, a_{ij}, b_i, b_{ij} \; \forall i = j = 1, 2$ are the coefficients.

In this configuration, there are two types of gene: one is known as the *control gene* and the other is called the *coefficient gene*. The control gene in bit form decides the state of activation for each block. "1" signifies the state of activation while "0" represents the state of de-activation. The coefficient gene (coded in either binary number or directly represented by a real number) defines the value of the coefficients in each block. The formulation can further be illustrated by an example as follows:

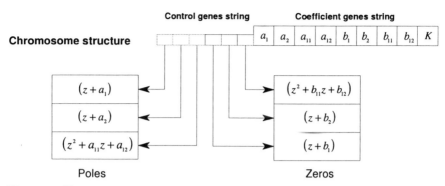

Fig. 5.2. Chromosome structure

Example 5.1.1. Consider a pulse transfer function:

$$H(z) = 1.5 \frac{z + 0.1}{z^2 + 0.7z + 0.8} \tag{5.4}$$

The formulation of the genes for a single chromosome, as indicated in Fig. 5.2, is thus

$$\underbrace{[0, 0, 1, 0, 1, 0,}_{Control\ genes} \underbrace{*^\dagger, *, 0.7, 0.8, *, 0.1, *, *, 1.5]}_{Coefficient\ genes} \tag{5.5}$$

In theory, such a chromosome may proceed to the genetic operation and each new generation of chromosomes can be considered as a potential solution to $H(z)$. However, the numerical values of genes may destabilize the filter if their values are not restricted. Since the filter is formed by a combination of first and second order models, the coefficients of the model can be confined within a stable region of an unit circle of z-plane. Hence, the coefficient of the first order model is simply limited to (-1,1), whereas the second order model needs a specific sub-routine to realize such a confinement. In this case, the coefficients of the denominator ($z^2 + a_1 z + a_2$) must lie within the stability triangle [198] as shown in Fig. 5.3.

In this way, the parameters of a_2 and a_1 can be initially set into their possible ranges, (-1,1) and (-2,2) respectively. If ($a_1 \leq -1 - a_2$) or ($a_1 \geq 1 + a_2$), then a_1 is assigned as a randomly generated number within the range ($-1 - a_2, 1 + a_2$). This process repeats for each newly generated chromosome, and the condition is checked to ensure that a stable filter is guaranteed. These confinements apply to the constraints of pole location of the filters for both cascade and parallel forms. These are classified and tabulated in Table 5.2.

---

† Don't care value

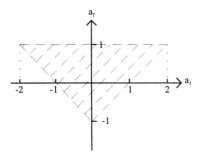

Fig. 5.3. Stability triangle for $(z^2 + a_1 z + a_2)$

Table 5.2. Constraints of pole locations

	Cascade Form	Parallel Form
Filter $H(z)$	$H(z) = K \prod_{i=1}^{n} \frac{(z+b_i)}{z+a_i} \times \prod_{i=1}^{m} \frac{(z^2+b_{j1}z+b_{j2})}{z^2+a_{j1}z+a_{j2}}$	$H(z) = \sum_{i=1}^{n} K_i \frac{(z+b_i)}{z+a_i} + \sum_{i=1}^{m} K_j \frac{(z^2+b_{j1}z+b_{j2})}{z^2+a_{j1}z+a_{j2}}$
Pole Constraints	$-1 < a_{j2} < 1$ $-1 < a_{j2} < 1$	$-1 - a_{j2} < a_{j1} < 1 + a_{j2}$

5.1.2 The Lowest Filter Order Criterion

Considering that the main purpose of control genes is to classify the topology of the filter structure, therefore, the length of this layer can be set to any length for the representation of a given filter. As this type of gene in the HGA formulation can be automatically turned "on" and "off" while proceeding with the genetic operation, a low order model may be obtained. In this way, those control genes signified "1" imply that an order(s) of the filter has been verified. As a result, the total number of "1"s being ignited would thus determine the final number of orders of the filter.

Based upon this principle, the lowest possible order of the filter may further be reached if a criterion is developed to govern this process. An extra objective function (f_3) has thus been proposed for this purpose. Given that a function f_3 is defined as the order of the filter, then by counting the "1"s within the control genes would be the final filter order of that particular chromosome, that is:

$$f_3 = \sum_{i=1}^{n} p_i + 2 \sum_{j=1}^{m} q_j \qquad (5.6)$$

where $(n + m)$ is the total length of control genes; p_i and q_j are the control bits governing the activation of i-th first order model and j-th second order model, although the maximum allowable filter order is equal to $(n + 2m)$.

5.1 Digital IIR Filter Design

In order to ensure f_3 is an optimal value for the filter design, an iterative method which is listed Table 5.3 can be adopted.

Table 5.3. Operation procedure

```
1. Define the basic filter structure used
2. Define the maximum number of iterations, N_max
3. Determine the value of n and m for H(z), (see Table 5.2)
4. The optimizing cycle is operated as follows
   While (tflag ≠ 1)
   {
      counter = 0;
      while ((sflag ≠ 1) and (tflag ≠ 1))
      {
         GAcycle(i,j);
         If ( f_1 = 0 and f_2 = 0 )
         {
            set n,m s.t. (n+2m) = p-1; /*p is the order of the solution*/
            sflag = 1;
         }
         else
            counter++;
         If counter= N_max
            tflag = 0;
      }
   }
```

Having now formulated the HGA for the filter design, a comprehensive design can be carried out for demonstrating the effectiveness of this approach. The parametric details of the design for the LP, HP, BP and BS filters are listed in Appendix E. For the purpose of comparison, all the genetic operational parameters were set exactly the same for each type of filter. The lowest order of each type of filter was obtained by HGA in this exercise. Table 5.4 summarizes the results obtained by this method and the final filter models can be stated as follows:

$$H_{LP}(z) = 0.0386 \frac{(z+0.6884)(z^2 - 0.0380z + 0.8732)}{(z-0.6580)(z^2 - 1.3628z + 0.7122)} \tag{5.7}$$

$$H_{HP}(z) = 0.1807 \frac{(z-0.4767)(z^2 + 0.9036z + 0.9136)}{(z+0.3963)(z^2 + 1.1948z + 0.6411)} \tag{5.8}$$

$$H_{BP}(z) = 0.077 \frac{(z-0.8846)(z+0.9033)(z^2 + 0.031z - 0.9788)}{(z-0.0592)(z+0.0497)(z^2 - 0.5505z + 0.5371)} \times$$

$$\frac{(z^2 - 0.0498z - 0.8964)}{(z^2 + 0.5551z + 0.5399)} \tag{5.9}$$

82   5. Genetic Algorithms in Filtering

$$H_{BS}(z) = 0.4919\frac{(z^2 + 0.4230z + 0.9915)(z^2 - 0.4412z + 0.9953)}{(z^2 + 0.5771z + 0.4872)(z^2 - 0.5897z + 0.4838)}$$
(5.10)

Table 5.4. Digital filters by HGA

Filter Type	Lowest Filter Order	Iteration	Objective vs Generation	Transfer Function	Response
LP	3	1649	Fig. 5.5a	H_{LP} (Eqn. 5.7)	Fig. 5.4a
HP	3	1105	Fig. 5.5b	H_{HP} (Eqn. 5.8)	Fig. 5.4b
BP	6	3698	Fig. 5.5c	H_{BP} (Eqn. 5.9)	Fig. 5.4c
BS	4	7987	Fig. 5.5d	H_{BS} (Eqn. 5.10)	Fig. 5.4d

It can seen that the HGA fulfils its promises as regards filter design. Judging from the frequency responses as indicated in Fig. 5.4, all the design criteria have been duly met. In addition, the HGA also provides another capability in which the lowest order for each filter can be independently obtained. This is undoubtedly due to the HGA's ability to solve multiobjective functions in a simultaneous fashion. Such a phenomenon is clearly demonstrated by the performance of objective functions, as indicated in Fig. 5.5, in which a lowest order filter is only obtained when the criteria for f_1, f_2 and f_3 are all simultaneously met. However, it should be noted that an adequate filter can be reached even when the functions f_1 and f_2 are fulfilled, while f_3 is not a minimum. This results in a higher order filter which sometimes is acceptable to meet the design specification and greatly shorten the computing effort.

To further highlight the improvement of digital filtering design by the HGA, this method has also been compared with classic filter designs, such as the BWTH, CHBY1, CHBY2 and ELTC approaches. Since the LP filter is the fundamental filter for the design of the other filters, a direct comparison with the LP filter only has been made. The results are summarized and tabulated in Table 5.5. The filter models were found as follows:

$$H_{BWTH} = \frac{0.0007z^6 + 0.0044z^5 + 0.0111z^4 + 0.0148z^3 + \cdots}{z^6 - 3.1836z^5 + 4.6222z^4 - 3.7795z^3 + \cdots}$$
$$\frac{\cdots + 0.0111z^2 + 0.0044z + 0.0007}{\cdots + 1.8136z^2 - 0.4800z + 0.0544}$$
(5.11)

$$H_{CHBY1} = \frac{0.0018z^4 + 0.0073z^3 + 0.0110z^2 + 0.0073z + 0.0018}{z^4 - 3.0543z^3 + 3.8290z^2 - 2.2925z + 0.5507}$$
(5.12)

$$H_{CHBY2} = \frac{0.1653z^4 - 0.1794z^3 + 0.2848z^2 - 0.1794z + 0.1653}{z^4 - 1.91278z^3 + 1.7263z^2 - 0.6980z + 0.1408}$$
(5.13)

5.1 Digital IIR Filter Design 83

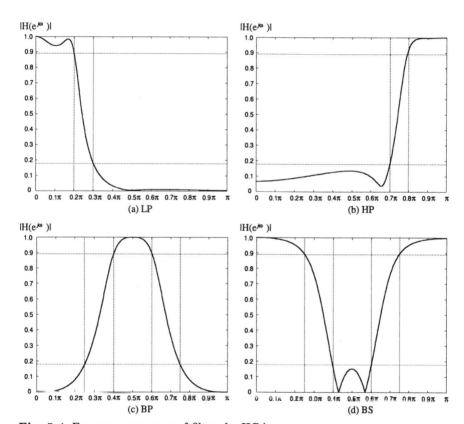

Fig. 5.4. Frequency response of filters by HGA

84   5. Genetic Algorithms in Filtering

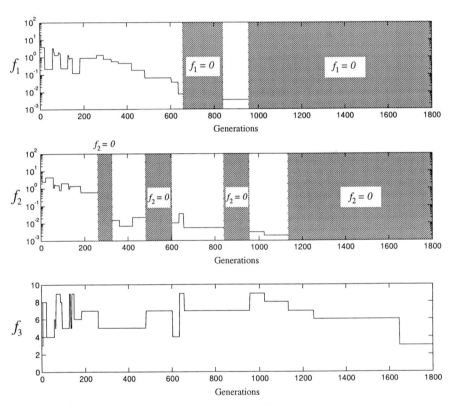

Fig. 5.5. Objective function vs generation for LP

$$H_{ELTC} = \frac{0.1214z^3 - 0.0511z^2 - 0.0511z + 0.1214}{z^3 - 2.1112z^2 + 1.7843z - 0.5325} \quad (5.14)$$

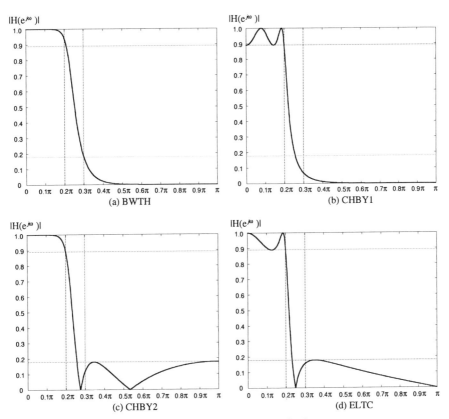

Fig. 5.6. Lowpass filter design using traditional methods

It is clearly shown from the frequency responses in Fig. 5.6 that the HGA is much superior to any of the classic methods. Only the ELTC method provides an equivalent order of the filter, but its frequency response is somewhat less attractive. A direct comparison of the magnitude of filter order is tabulated in Table 5.6. It is clearly indicated that the method of HGA for digital filtering design is not only capable of reaching the lowest possible order of the filter, but its ability to satisfy a number of objective functions is also guaranteed. It should also be noted that, the design of HP, BP and BS filters can be independently assigned by the method of HGA, which is a unique asset for designing filters that involve complicated constraints and design requirements.

Table 5.5. Result of various filters comparison

LP Filter	Lowest Filter Order	Transfer Function	Response
BWTH	6	Eqn. 5.11	Fig. 5.6a
CHBY1	4	Eqn. 5.12	Fig. 5.6b
CHBY2	4	Eqn. 5.13	Fig. 5.6c
ELTC	3	Eqn. 5.14	Fig. 5.6d
HGA	3	Eqn. 5.7	Fig. 5.4a

Table 5.6. Lowest filter order due to various design schemes

Filter	BWTH	CHBY1	CHBY2	ELTC	HGA
LP	6	4	4	3	3
HP	6	4	4	3	3
BP	12	8	8	6	6
BS	12	8	8	6	4

5.2 Time Delay Estimation

In this section, a GA is applied to tackle an on-line time-delay estimation (TDE) problem. TDE can be found in many signal processing applications such as sonar, radar, noise cancellation, etc. It is usually solved by changing the one-dimensional delay problem into a multidimensional problem with finite impulse response (FIR) filter modelling [26, 177, 241]. Adaptive filtering techniques [101, 236] have been successfully applied in this area relying on the unimodal property of FIR error surface. Due to the GA's ability in locating global optima on a multimodal error surface, direct estimates of delay as well as the gain parameter are possible.

5.2.1 Problem Formulation

Estimation of the time delay between signals received from two sensors is considered. Eqn. 5.15 represents the conventional discrete time delay model.

$$x(kT) = s(kT) + v_1(kT) \quad \text{and}$$
$$y(kT) = \alpha s(kT - D) + v_2(kT) \tag{5.15}$$

where T is the sampling period; $x(kT)$ and $y(kT)$ are two observed signals at sampled time kT; $s(kT)$ is the transmitted source, white signal with power σ_s^2; D is the delay to be estimated; α is a gain factor between the sensors; and $v_1(kT)$ and $v_2(kT)$ are zero-mean stationary noise processes with power σ_n^2, assumed to be uncorrelated with each as well as $s(kT)$.

In principle, the searching domain of time delay includes infinite numbers of values of D and α. In practice, these numbers are limited by the knowledge of the delay and gain range and by the desired resolution. Moreover, the resolution of delay is often much finer than the sampling interval T. Assuming that the measured signals are band-limited with a frequency range $(-\omega, \omega)$ which leads to $T \leq (2\omega)^{-1}$, the signal $x(kT)$ can be interpolated in the form of sinc function [26]. Therefore, the delayed version $x(kT - D)$ can be approximated by

$$x(kT - D) \approx \sum_{i=-L}^{L} sinc(i - D) x(k - i) \quad (5.16)$$

The estimation error, $e(k)$, is defined as

$$e(k) = y(k) - AX(k) \quad (5.17)$$

where $X(k) = \begin{bmatrix} x(k+L) & x(k+L-1) & \cdots & x(k-L) \end{bmatrix}^T$ is the input vector, and A is the vector to be optimized.

The optimal solution of A ($\tilde{A}$) for minimum mean square error (MMSE) criterion is expressed as

$$\tilde{A} = \frac{\alpha \sigma_s^2}{\sigma_s^2 + \sigma_n^2} \begin{bmatrix} sinc(-L - D) & sinc(-L + 1 - D) \\ & \cdots & sinc(L - D) \end{bmatrix} \quad (5.18)$$

which is only dependent on α and D. Hence, the problem is now reduced to two parameters $(g, \hat{D})$:

$$g = \frac{\alpha SNR}{1 + SNR} \quad \text{and} \quad \hat{D} = D \quad (5.19)$$

where SNR is the signal to noise ratio.

The optimal set of $(g, \hat{D})$ lies upon the multimodal mean square error surface which might not be easily obtainable using gradient searching methods. Hence, optimization of A vector leads to the result of an FIR filter. However, such a filtering model often causes estimation noise problem as more parameters are required for estimation. To combat this difficulty, the Constrained LMS Algorithm [203] to search for $\hat{D}$ directly with the assumption that the initial guess $\hat{D}(0)$ is within the range of $D \pm 1$ has been proposed.

5.2.2 Genetic Approach

To directly estimate the optimal set of $(g, \hat{D})$ and avoid any restriction of the initialization, a GA can be introduced. The associated delay and gain in this case, directly represented by a binary string, i.e. chromosome, without

5. Genetic Algorithms in Filtering

modelling of the delay is a significant advantage. This chromosome is defined in Eqn 5.20 and its structure is depicted in Fig. 5.7.

$$I = (b_1, b_2, \ldots, b_{32}) \tag{5.20}$$

where $b_i \in B = 0, 1$

Fig. 5.7. Chromosome structure

Consider a transformation from a binary to a real number, $\Lambda : B^{16} \times B^{16} \longrightarrow \Omega \subseteq \Re \times \Re$ converting the chromosome bit string $I = (b_1, b_2, \ldots, b_{32})$ to phenotype values $(g, \hat{D})$. The overall goal of GA for time delay problems is to obtain the chromosome $I^* \in B^{16} \times B^{16}$ with the phenotype value $\Lambda(I^*)$ such that

$$\Psi(\Lambda(I)) \geq \Psi(\Lambda(I^*)) = \Psi^* \qquad \forall I \in \Omega \tag{5.21}$$

where $\Psi : \Omega \longrightarrow \Re^+$ is the average square difference function expressing as $\Psi(g, \hat{D}) = \frac{1}{N} \sum_{k=k_0}^{k_0+N-1} \left[y(k) - g \sum_{i=-L}^{L} sinc(i - \hat{D}) x(k-i) \right]^2$; Ω is the searching domain for $(g, \hat{D})$; N is the estimation window size; k_0 is the starting sample of the window; Ψ is called a global minimum, and $\Lambda(I^*) = (g^*, \hat{D}^*)$ is the minimum location in the searching space Ω.

In order to apply a GA for real time TDE applications as stated in Eqns. 5.15 to 5.21, the problem on noise immunity and robustness should be addressed.

To tackle the system noise immunity problem, that is when the noise level is comparative to the actual signal, i.e. a low SNR value, the obtained result is not steady and often yields high variance. In order to improve the noise immunity, the phenotype values for the best chromosomes obtained in the generations are stored in particular memory locations. The mean of these phenotype values is applied for real world interaction instead of the phenotype value of the current best chromosome.

Furthermore, in order to reach a global optimum, a strong selection mechanism and a small mutation rate are often adopted during the GA operation, in order to contract diversity from the population through the searching process. Should the environmental condition be changed, the conventional GA is unable to redirect its search to the new optimum speedily. In this system, a statistical monitoring mechanism is necessary and implemented.

By monitoring the variance and the mean of the output data, a change of environment can be detected. It is assumed that the process is subject only to its natural variability and remains in a state of statistical control unless a special event occurs. If an observation exceeds the threshold, a significant deviation from the normal operation is deemed to have occurred. The GA operations have to be readjusted to adapt changes.

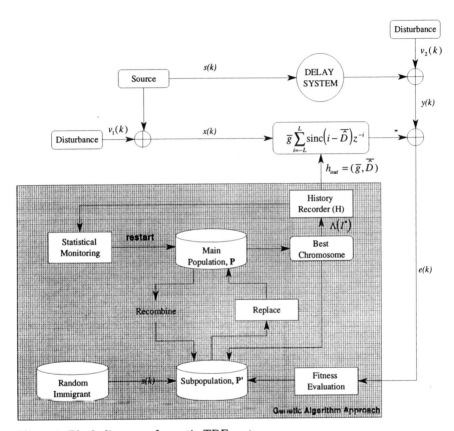

Fig. 5.8. Block diagram of genetic-TDE system

By taking all these considerations into account, a genetic-based time delay estimation system is thus proposed and schematically depicted in Fig. 5.8. Basically the system has three modes of operation, namely, Initialization, Learning and Monitoring. The procedures for each mode are described as follows:

Initialization Mode. The activation of the TDE system should first be started with initialization procedures as follows:

90   5. Genetic Algorithms in Filtering

1. Generate the chromosomes, I, as stated in Eqn. 5.20 for the main population (P) with uniform distribution;
2. Evaluate the objective value, Ψ, as in Eqn. 5.21 for each chromosome from population P

Learning Mode. Once the initialization mode has been completed, the GA operations can proceed. This Learning Mode enables chromosomes to be improved while the system is in operation. The population is updated recursively, and always supplies the mean of the past best phenotype values for real world operation. The learning cycle is depicted in Fig. 5.9.

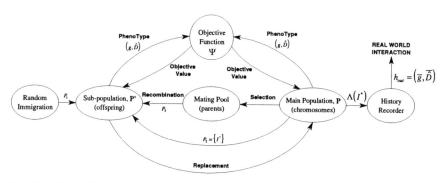

Fig. 5.9. Learning cycle

The following GA operations are used to ensure the speed of computation as well as to improve the noise immunity within the learning cycle.

Parent Selection. Parent Selection is a routine to emulate the survival-of-the-fittest mechanism of nature. It is expected that a better chromosome will receive a higher number of offspring and thus have a higher chance of surviving in the subsequent generation. The chromosomes in the population pool are selected for the generation of new chromosomes (offspring) by the stochastic universal sampling (SUS) method [10].

Formulation of Sub-population (P'). The new chromosomes in the sub-population (P') are generated by combining three different sets:

$$P'_k = P_{1,k} \cap P_{2,k} \cap P_{3,k} \tag{5.22}$$

where $P_{1,k}$ is the set of chromosomes to be generated via random immigration; $P_{2,k} = I_k^*$ contains the best chromosome in the current main population; and $P_{3,k}$ is the set of chromosomes generated through a recombination process.

Genetic Operation. 2-point crossover and bit mutation are applied.

Replacement. In order that a best chromosome is available for real-world interaction, the main population is discarded and replaced by the newly-obtained subpopulation (P') in the next generation. That is when

$$P_{k+1} = P'_k \tag{5.23}$$

Fitness Assignment. To prevent premature convergence, the fitness value of the chromosome, $f(I,k)$, is assigned by a linear ranking scheme [9].

History Recorder. This scheme is necessary in order to maintain the best chromosome of the current population for real-time interaction. The History Recorder H, is structured as

$$\begin{aligned} h(i) &= h(i+1) \quad \forall\, i \in (0, N) \\ h(N) &= \Lambda(I^*_k) \end{aligned} \tag{5.24}$$

where I^*_k is the best chromosome in the k-th generation; N is the maximum size of the History Recorder; and $h(i)$ is the i-th element in History Recorder.

To further improve the system noise immunity capability, the mean value of the elements in the History Recorder, (h_{out}), for real world operation is adopted:

$$h_{out} = \frac{1}{N} \sum_{i=1}^{N} h(i) \tag{5.25}$$

Monitoring Mode. Under normal circumstances, the TDE system should properly operate in a learning mode. Its performance should thus be improved and enhanced further as time continues. However, this mode of operation must be guaranteed once the system is disturbed. Therefore, in order to increase the system's robustness, since it has the capacity to reject external disturbance, a monitoring scheme is set-up in parallel for the on-line checking of the environment. This can be established by using the statistical output value to provide the necessary information about a change of situation. The scheme is formulated by considering the set $S \subset II$, which consists of n output values from current sample (k) as to the past sample ($k - n - 1$), and the system is thus considered as being converged or unchanged when the standard deviation is less than a particular value σ_k. This value can be obtained either experimentally or where a prior knowledge is available.

$$\sigma(S) < \sigma_k \tag{5.26}$$

The system should only be re-set when

$$\begin{aligned} |\hat{D} - \bar{D}| &> \delta_D \quad \text{or} \\ |g - \bar{g}| &> \delta_g \end{aligned} \tag{5.27}$$

where $(g, \hat{D})$ is the current output value; $(\bar{g}, \bar{D})$ is the mean value after convergence; δ_D, δ_g are the threshold values which can be determined as the multiple of the σ_k or the expected accuracy of the system.

5.2.3 Results

To illustrate the effectiveness of the proposed method for TDE, both time invariant and variant delay cases were studied. In the case of invariant time delay, only the estimation of the time delay element was considered, whereas for time variant delay, the introduction of both gain and delay parameters were given. To further ensure the tracking capability and robustness of the system, which was generally governed by the monitoring scheme, the previous experimental run was repeated again while the monitoring mode was being switched off. A direct comparison of the results obtained from both cases was made to assess the function of this mode.

Invariant Time Delay. For the estimation of time-invariant time delay, the gain was fixed to unity and the sampling time $T = 1$. Since only the time delay was to be estimated, the chromosome was formulated as a 16-bit string which solely represented the delay, see Fig. 5.20. The parameters used for simulation runs were $\sigma_s^2 = 1$, $L = 21$, $D = 1.7$ and $N_p = 8$. The genetic operations consisted of a two-point crossover with an operation rate $p_c = 0.8$ and a mutation with a rate $p_m = 0.01$. Objective Function defined in Eqn 5.21 was evaluated while the window size was arbitrarily limited to five samples. Figs. 5.10 and 5.11 show the comparison with LMSTDE [236] and the Constrained LMS Algorithm [203] whose formulations are stated in Appendix A and B for both noiseless and noisy condition $(SNR = 0dB)$. For the noiseless situation, a direct output of the current best chromosome was applied.

The mean and variance of different algorithms are compared in Table 5.7. It can be observed that the proposed method provides a better delay estimation as compared with the two other methods.

Table 5.7. Statistical comparison of different algorithms

Algorithms	Mean	Variance (10^{-4})
Proposed Algorithm	1.6982	0.569
Constrained LMS Algorithm	1.6827	7.714
LMSTDE	1.7045	2.584

Variant Time Delay. In order to demonstrate the robustness of the proposed system, an estimation of time variant delay is also considered. The sequences $s(kT)$, $v_1(kT)$ and $v_2(kT)$ were generated by a random

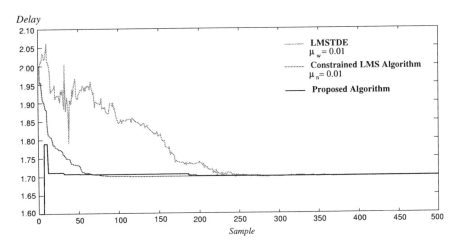

Fig. 5.10. Comparison of different algorithms (noiseless condition)

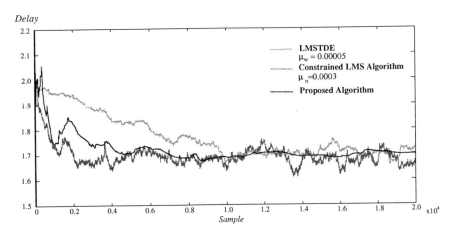

Fig. 5.11. Comparison of different algorithms ($SNR = 0dB$)

94   5. Genetic Algorithms in Filtering

number generator of Gaussian distribution with the signal power σ_s^2 set to unity. The tracking ability of the proposed method for noiseless condition is demonstrated in Fig. 5.12. A step change on the parameters of gain and delay for the TDE system occurred in every 1,000 samples.

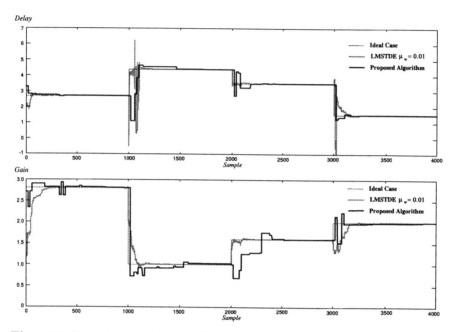

Fig. 5.12. Time delay tracking by different algorithms (noiseless condition)

For the noiseless case, the current best chromosome is considered as the output. It should be noted that the Constrained LMS Algorithm is not applicable for time variant delay problems, as an initial guess is not easily determined within the required limit, $D \pm 1$. Hence, comparison is only possible via the traditional LMSTDE method.

Fig. 5.13 shows the results obtained for the case when SNR=0dB is set. Both variation of gain and delay changes remained unchanged as from previous cases. It can be seen that the system behaves well under extremely noisy conditions.

To illustrate the effectiveness of the monitoring system, the previous simulation run was repeated once again, but the monitoring mode was switched off. The tracking ability of the system was poor as indicated in Fig. 5.14. It can be observed that the system reacts very slowly to the changing environment in the absence of the monitoring scheme.

5.2 Time Delay Estimation

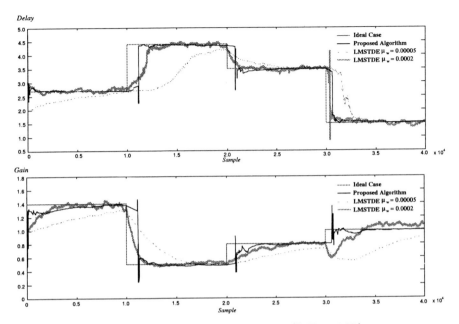

Fig. 5.13. Time delay tracking different algorithms (SNR = 0dB)

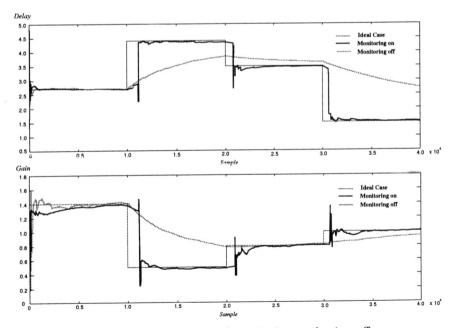

Fig. 5.14. Slow tracking when statistical monitoring mechanism off

5.3 Active Noise Control

The concept of noise reduction using active noise control (ANC) has been with us for a long time. The very first patent design by Leug [131] was established in 1936. ANC is a technique that uses secondary acoustic sources to generate the anti-phase sound waves that are necessary for cancelling undesired noise. It works on the basic principle of destructive interference of sound fields. A number of ANC systems have been developed for different applications [57]. The more notable systems are: ANC systems for power transformers [20, 32], duct systems [93, 166], and systems for vehicle or flight cabins [56, 207]. The main contribution to the success of these systems is the use of parameter identification techniques for the estimation of the noise dynamics and the development of adaptive optimal control laws to govern the corresponding anti-phase acoustic signals for final noise cancellation. An effective ANC system must be able to adjust the controller to activate the secondary sources in such a way that the resultant noise received by the error sensor is reduced to a minimum.

Therefore, speedy and accurate estimates of the acoustic dynamics are the keys elements to the success of ANC systems. This generally involves the determination of the correct estimates of the amplitude and phase of noise signals. Thus far, the well known gradient climbing type [74, 236] of optimization scheme has normally been adopted for this purpose. However, this process is not without its shortcomings. The most noticeable adverse effect is the inherent phenomenon of being trapped in the multimodal surface during the process of optimization. Another problem is the structure of parallelism for high speed computation which is considered an essential technique for the development of complex and multichannel systems.

In this chapter, the ANC system adopts the intrinsic properties of the GA. A number of new schemes are proposed in this development. These include the modelling of acoustic path dynamics, the parallelism of computation and a noticeable multiobjective design for sound field optimization. An advanced ANC design that uses dedicated hardware for GA computation is also recommended. The advantage of developing this architecture is its simplicity in system integration, and the fact that an expensive DSP processor is no longer required for calculating the time consuming numerical values.

5.3.1 Problem Formulation

To configure an ANC system based on a GA formulation, the overall feedforward ANC system can be seen from Fig. 5.15. The aim is to minimize the sound field at "D" via "C" for generating the anti-phase noise signal to counteract the noise produced by "A". This is only possible if "B", the detector (microphone) is placed adjacent to "A" and sends its output signal to the controller $C(z)$ to adjust the outgoing noise by the actuators (loudspeakers). An optimal noise reduction performance would only be on hand if another microphone(s) is located at "D" and feeds its output signal to the controller $C(z)$ via some intelligent optimization routines.

In practice, other adverse conditions can be contributed by the locations of the microphones and the loudspeakers. A positive acoustic feedback path always exists due to a contaminated secondary signal sensed by the microphone at "B", i.e. through location "C" to "B". This path introduces an additional dynamic into the overall noise dynamics. The other transfer function relating to the loudspeaker and the error sensor, i.e. through location "C" to "D", is non-unity, time-varying and unknown. A slight positional mismatch between the loudspeakers and the error sensor will cause significant deterioration in the performance of ANC systems. Therefore, it is necessary to take these problems into account while the identification procedure is taking place. Generally, this process can be completed by the well known filtered-x Least Mean Squares [58, 166].

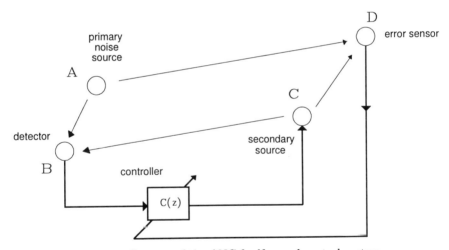

Fig. 5.15. Schematic diagram of the ANC feedforward control system

Mathematically, the problem can be formulated according to the general block diagram of the ANC system as depicted in Fig. 5.16. In order that $o(k)$ receives minimal energy due to $s(k)$ propagation, $u(k)$ has to generate appropriate anti-phase signals, such that

$$o(k) = H_3(z)s(k) + H_4(z)u(k) \quad (5.28)$$
$$m(k) = H_1(z)s(k) + H_2(z)u(k) \quad (5.29)$$

Substituting Eqn. 5.29 into Eqn. 5.28, we have,

$$o(k) = H_3(k)\frac{m(k) - H_2(z)u(k)}{H_1(z)} + H_4(z)u(k) \qquad (5.30)$$

Eqn. 5.30 is a global formulation of all ANC configurations including its application in a 3-D propagation medium.

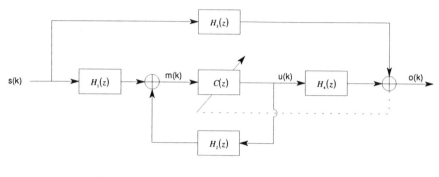

$s(k)$: primary source signal
$u(k)$: secondary source signal
$m(k)$: detector signal
$o(k)$: error signal
$C(z)$: transfer function of the controller
$H_1(z)$: transfer function of the acoustic path through primary source to detector
$H_2(z)$: transfer function of the acoustic path through secondary source to dectector
$H_3(z)$: transfer function of the acoustic path through primary source to error sensor
$H_4(z)$: transfer function of the acoustic path through secondary source to error sensor

Fig. 5.16. General block diagram of adaptive active noise attenuation system with input contamination

In general, the following assumptions must be held.
Assumptions :

(A1) $H_1(z)$, $H_2(z)$, $H_3(z)$ and $H_4(z)$ are stable rational transfer functions (by finite dimensionality) of acoustic paths.
(A2) The number of pure delays of $H_1(z)H_4(z)$ is less than or equal to that of $H_3(z)$, i.e. $\tau_1 + \tau_4 \leq \tau_3$.

The assumption (A1) is necessary and enables the acoustic systems be identified by the use of parameter identification techniques. In addition, a causal system is obtained as long as the assumption (A2) holds.

Design Procedure. A good ANC performance can be obtained if the parameters of the transfer functions of $H_1(z)$, $H_2(z)$, $H_3(z)$ and $H_4(z)$ are accurately identified so that an optimum controller $C(z)$ is derived to drive the secondary sources to compensate for the noise interference. The operational procedure of the active noise control system can be divided into two stages. The first stage is to optimize the transfer function model for the acoustic paths whereas the second stage is to optimize the transfer function model of the controller $C(z)$.

To further enhance the system's stability, the troublesome positive acoustic feedback path $H_2(z)$ from the secondary loudspeakers to the detector must be eliminated. This is possible by the use of a piezoelectric accelerometer as the primary source detection device. This sensor picks-up only the mechanical vibration rather than the actual acoustic signal, although a direct relationship between the vibration forcing signal and the acoustic sound pressure waves can be established. In this way, the accelerometer senses only the noise source vibration signal and is unaware of the acoustic signal due to secondary sources (loudspeakers). To formulate the controller $C(z)$ for noise control, the procedure is much simplified and is listed as follows:

Step 1: Estimation of the transfer function from the detector to the error sensor, $H_3(z)H_1^{-1}(z)$
Recall Eqn. 5.30 and $H_2(z) = 0$, the resultant signal at the error sensor can be expressed as

$$o(k) = H_3(k)\frac{m(k)}{H_1(z)} + H_4(z)u(k) \tag{5.31}$$

Consider an interval $0 < k \leq N_1$ while the secondary source is turned off. Here

$$u(k) = 0 \quad \text{and} \quad u(k)H_4(z) = 0 \tag{5.32}$$

Substitute Eqn. 5.32 into 5.31, the error signal is thus

$$o(k) = m(k)H_3(z)H_1^{-1}(z) \tag{5.33}$$

Consider $m(t)$ and $o(t)$ as the input and output of the transfer function, $H_3(z)H_1^{-1}(z)$, this step is equivalent to parameter identification of the unknown transfer function $H_3(z)H_1^{-1}(z)$.

Step 2: Estimation of the controller, $C(z)$ To minimize $o(k)$ as indicated in Eqn. 5.31, for the optimal noise cancellation, then

$$-m(k)H_3(z)H_1^{-1}(z) = [m(k)H_4(z)]\,C(z) \tag{5.34}$$

Similarly, the controller design for $C(z)$ can also be considered a parameter identification problem while $m(k)H_4(z)$ is available. In this case, $m(k)H_4(z)$ is obtained by the following procedure:

Consider $N_1 < k \leq N_1 + N_2$ while the reference signal $m(k)$ is transmitted through the secondary sources, that is,

$$u(k) = m(k) \tag{5.35}$$

Then, from Eqn. 5.31,

$$m(k)H_4(z) = o(k) - m(k)H_3(z)H_1^{-1}(z) \tag{5.36}$$

Since $H_3(z)H_1^{-1}(z)$ has already been obtained in the previous step, $m(k)H_4(z)$ can thus be obtained by Eqn. 5.36.

Estimation Model. Different kinds of filter models can be adopted for the ANC problem. Due to the limitation of the gradient climbing technique, the most common model applied is the FIR filter. Since a Digital Signal Processor (DSP) is used for real time ANC, the digital FIR model is normally used,

$$H(z) = \sum_{i=0}^{L-1} a_i z^{-i} \tag{5.37}$$

where L is the filter length, and a_i is the filter coefficient.

Considering that the ANC configuration is largely affected by the inherent nature of time delay, it is obvious that a large portion of the high order filter order is being using for time delay modelling. Hence, the inclusion of the time delay element in the model should prove a more appropriate approach instead of the usual high order filter.

Eqn. 5.38 proposes a general form of the modified model for the estimation of the acoustic path process [219].

$$H(z) = gz^{-d} \sum_{i=0}^{L-1} b_i z^{-ni} \tag{5.38}$$

where g is the appropriate d.c gain; d is the time delay element; L is the number of tap; and n is the tap separation.

An evaluation process for verification of this model was conducted in terms of acoustic path estimation. Results were compared for the conventional 81-tapped* FIR filter model and the modified 21-tapped[†] FIR filter model with $n = 4$ in Eqn. 5.38. Double tone noises of frequencies 100Hz and 250Hz were applied for this evaluation exercise. The residue error signals $o(k)$ due to the use of these filters are shown in Fig. 5.17. It is evident that the lower order modified FIR model is far better than the conventional higher order FIR filter. The time response is fast and high frequency filtering is also present, and this is considered to be an important asset for ANC. Judging from these results, the inclusion of delay and gain in the model will enhance the noise reduction performance.

---

* For conventional FIR, one tap is defined as $\left[b_i z^{-i}\right]$ where i is an integer.
[†] For modified FIR, one tap is defined as $\left[b_i z^{-ni}\right]$ where i and n are both integers. $n = 4$ is experimentally determined.

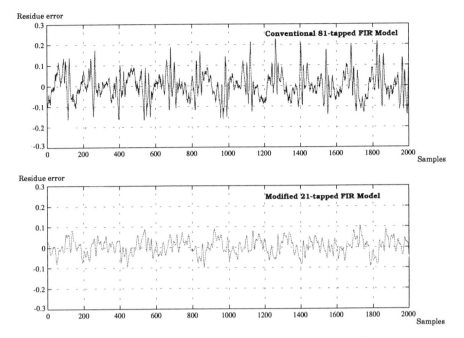

Fig. 5.17. Comparison of the conventional and modified FIR models

It should be noted that the traditional gradient technique is ill suited to estimate this modified model correctly, particularly for the variables of gain and delay elements, which are initially unknown and belong to the class of multimodal error surfaces for optimization.

5.3.2 Simple Genetic Algorithm

Parameter Identification. As derived in Sect. 5.3.1, ANC is now converged into a two-stages parameter identification problem. Fig. 5.18 depicts the use of GA for such a problem.

The objective is to minimize the error between the unknown system output and the output of the estimation model. Hence, the objective function may be defined as the windowed mean square error

$$f = \frac{1}{N} \sum_{i=1}^{N} (y(k) - x(k)H(z))^2 \qquad (5.39)$$

where $x(k)$ and $y(k)$ are the digitized input and output values of the unknown system, respectively; and N is the window size.

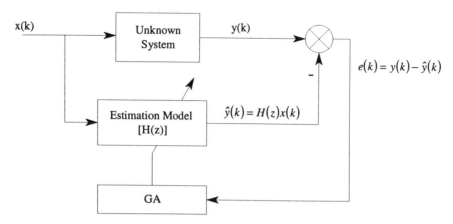

Fig. 5.18. Parameter identification using GA

Genetic Algorithm for Parameter Identification. Based on the new model expressed in Eqn. 5.38, the structure of the chromosome is defined and formulated as follows:

$$I = \{d, g, B\} \in \Phi \subset Z \times \Re \times \Re^L \tag{5.40}$$

where $d \in [0, \alpha] \subset Z$; $g \in [0, \beta] \subset \Re$; and $B = [b_0, b_1, \ldots, b_{L-1}] \in [-1, 1]^L \subset \Re^L$ with Z and $\Re$ are the sets of integer and real numbers, respectively; α and β are the maximum range of d and g, respectively.

It should be stressed that this constrained model introduces a multimodal error surface which is considered impossible for the application of traditional gradient methods. On the other hand, GA guarantees that the objective function is globally optimized despite its multimodal nature and the constraints of the model. A brief summary of the GA operations required to achieve the optimization goal is given below:

Fitness Assignment. The fitness value of the chromosome is assigned by a linear ranking scheme [234] based on the objective function in Eqn. 5.39.

Parent Selection. The stochastic universal sampling (SUS) method [10] is applied for selection.

Crossover. The genes of the chromosome can be classified as two types: delay-gain genes $[d, g]$ and filter coefficient genes $[b_0, b_1, \ldots, b_{L-1}]$. A one point crossover operation [78] was applied to both type of genes independently.

Mutation. In natural evolution, mutation is a random process designed to introduce variations into a particular chromosome. Since the genes for delay and gain $[d, g]$ as well as those for filter coefficients are represented by integer or real numbers, random mutation [149] was applied.

Insertion Scheme. The newly-generated chromosome is reinserted into the population pool if its fitness value is better than the worst one in the population pool [78].

Termination Criterion. Since GA is a stochastic searching technique, it experiences a high variance in response time. Hence, the progress per generation is used to determine the termination of the GA.

$$t = \begin{cases} 1 & \text{if } \Gamma_k = \Gamma_{k+1} \quad \forall i \in (0, r] \\ 0 & \text{otherwise} \end{cases} \quad (5.41)$$

where Γ is the population at k-th generation.

Table 5.8. Relationship of r, objective value and termination generation

r	$(\bar{f})$	$(\bar{\tau})$
2	12.2451	161.05
4	1.3177	334.15
6	0.8662	414.60
8	0.5668	512.20
10	0.5283	579.30
12	0.5076	670.50
14	0.4894	742.90
16	0.4866	827.30
18	0.4792	933.85
20	0.4662	984.30

In order to determine a proper value of r, 20 experimental trials were conducted. The relationship between the objective mean value $(\bar{f})$, and the terminated generation mean $(\bar{\tau})$ is tabulated in Table 5.8. A trade-off between the two values is made in selecting r. If the value of $r > 8$ is selected, a slight improvement in accuracy may be achieved but the real-time performance deteriorates. Hence, the empirical data indicated that $r = 8$ was a reasonable choice for terminating the GA production.

Implementation. Although the above learning procedure of the GA can achieve the parameter identification in an ANC problem, to avoid the intrinsic problem of randomness of the GA at the initial stage, and to guarantee at least some level of noise reduction at the beginning, it is necessary to combine the GA and the traditional gradient techniques in an efficient manner to achieve the ANC objective. The traditional gradient technique may not provide the required optimal performance, but its instantaneous response is an asset to the human hearing response as well as to real time control. Therefore, both GA and traditional gradient optimization procedures should be integrated for this purpose. Initially, the performance of the system using a low order FIR filter with traditional gradient optimization routines need not be optimal, and even the level of noise reduction may be low. The

104   5. Genetic Algorithms in Filtering

controller $C(z)$ will be continuously updated when a global solution is found by the GA for the modified model. This can only be realized by hardware via a communication link between these two processes. In this way, the real-time deadline will be met and an optimal noise control performance is also guaranteed. Fig. 5.19 shows the parallel hardware architecture, using two independent TMS320C30 digital signal processors for such implementation [219].

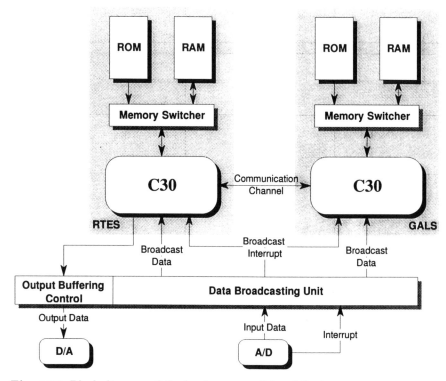

Fig. 5.19. Block diagram of the hardware parallel architecture

The architecture consists of two individual units, known as the Real-Time Executed System (RTES) and the Genetic Algorithm Learning System (GALS). RTES is used to provide a speedy, but controllable, solution of the system by conventional Recursive Least Squares (RLS) while GALS optimizes and refines the controller, in order to achieve the required optimal noise control performance. Each system is implemented using a TMS320C30 processor together with its own local memory. To prevent the data access serialization and delays that are usually experienced by each unit, a data broadcasting device was designed to handle the distribution of the external data addressed by each unit. Such a broadcasting unit releases the dependence of each processor since each one virtually has its own input device. As a result, the inter-units communication command is greatly minimized.

Statistic Monitoring Process. As the system is designed to adapt to the change of noise environment, the run-time primary source signal $m(k)$ must be monitored in order to guarantee the performance requirement. This signal is compared with the estimated signal $\hat{m}(k)$ to confirm whether there is any change of environment taking place. $\hat{m}(k)$ is calculated by

$$\hat{m}(k) = g_m \sum_{i=0}^{L-1} [\alpha_i m(k - 1 - ni - d_m)] \qquad (5.42)$$

The parameters α_i and g_m are learned by the GA using the data sequence of $m(k)$ collected in the past optimization process. Hence, the estimated error $e(k)$ is expressed as:

$$e(k) = m(k) - g_m \sum_{i=0}^{L-1} [\alpha_i m(k - 1 - ni - d_m)] \qquad (5.43)$$

The mean ($\bar{e}$) and variance σ^2 of the $e(k)$ within the data sequence can thus be determined. A Statistical Control procedure was established to ensure the robustness of this scheme. This assumes that the process is subject only to its natural variability and remains in a state of statistical control unless a special event occurs. If an observation exceeds the control limits, a statistically significant deviation from the normal operation is deemed to have occurred, that is when:

$$m(k) - \hat{m}(k) > \bar{e} \pm 3\sigma \qquad (5.44)$$

Any change of environment will cause the restart of the RTES learning cycle automatically.

Experimental Setup and Results. The performance of the system was investigated using specifically-designed experimental equipment to realize the active noise control configuration shown in Fig. 5.20. It comprises a primary source (loudspeaker) and four additional secondary sources which were located close to the primary source with a quadpole arrangement, using four small loudspeakers. Fig. 5.20 shows the quadpole arrangement of the primary and secondary sources. The circle indicated with the mark '+' denotes the primary noise source and the other circles with the marks '-' denote the secondary sound sources [123].

The error microphone was placed perpendicular to the vertical plane of the primary and secondary sources at a distance of about 1m away from the centre of the primary source. This meant that the position of the error microphone could be in the doublet plane of symmetry in order to obtain an

5. Genetic Algorithms in Filtering

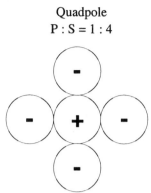

Quadpole
P : S = 1 : 4

Fig. 5.20. Geometry of the primary sound source 'P' (+) and the secondary sound sources 'S' (-)

optimal performance [93]. The piezoelectric accelerometer was attached to the primary source.

The experiments were conducted in a general laboratory with a dual tone noise signal of 100Hz and 250Hz. The sampling frequency for the Analog-to-Digital Converter was 10kHz.

The parameters of the subsystems were set as below:

- RTES - $H_3(z)H_1^{-1}(z)$ and $C(z)$ were modelled by traditional 21-tapped FIR filters and 1000 iterations of the Recursive Least Square algorithm were used to estimate the coefficient values.

- GALS - $H_3(z)H_1^{-1}(z)$ and $C(z)$ were estimated in the form of Eqn. 5.38 with a delay parameter (d) and a gain factor (g) for the modified 21-tapped FIR filter. The searching space was defined as below:

$$d \in [0, 100] \subset Z$$
$$g \in [0, 5] \subset \Re$$
$$B = [b_0, b_1, \ldots, b_{20}] \in [-1, 1]^{21} \subset \Re^{21}$$

The experimental noise level was recorded by a RION 1/3 Octave Band Real-Time Analyzer SA-27. Table 5.9 shows the power sum levels of all the bands and the power sum level with a frequency A-weighted characteristic when the ANC system is being turned on and off. The results are depicted in Fig. 5.21.

It can be seen from these results that the GA operated scheme GALS outperforms the conventional FIR filters in RTES. In addition, the dual tone frequency signals are greatly reduced by more than 15dB each. The high frequency noise is also suppressed. With the RTES scheme, this phenomenon

5.3 Active Noise Control

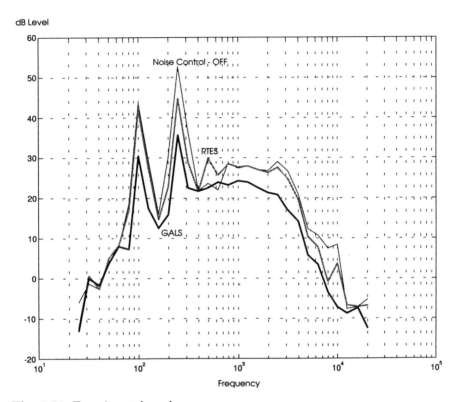

Fig. 5.21. Experimental result

Table 5.9. Power sum value for ANC system on and off

	A_p/dB	$A_p(\omega)/dB$
ANC – OFF	63.9	53.5
RTES – ON	61.9	47.1
GALS – ON	52.9	38.4

was not observed and the general noise reduction performance was also very poor when using the equivalent low order FIR filters.

5.3.3 Multiobjective Genetic Algorithm Approach

One of the distinct advantages of the GA is its capacity to solve multiobjective functions, and yet it does not require extra effort to manipulate the GA structure in order to reach this goal. Therefore, the use of this approach for ANC makes it a very good proposition to optimize a "quiet zone" and, at the same time, alleviates the problem of selecting the error sensor (microphones) placement positions at the quiet end to achieve a good result.

Consider a multiple channel ANC system that consists of m error sensors and n secondary sources in an enclosure depicted in Fig. 5.22. The GA can be used to optimize each error sensor independently to fulfil their targets.

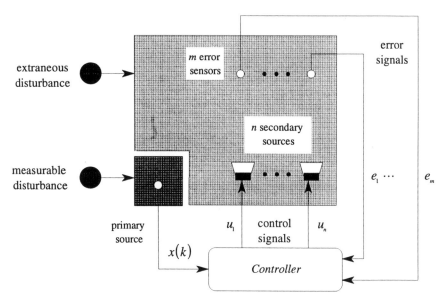

Fig. 5.22. Multiple channel ANC system

Multi-objective Functions. At error sensor i, an objective function (f_i) is defined as below:

$$f_i = \frac{1}{N} \sum_{k=0}^{N-1} e_i^2(k) \qquad (5.45)$$

where $e_i(k)$ is the acoustic signal obtained at error microphone i at time k; and N is the window size.

Instead of forming a single criterion function by lumping the objective functions with a linear or nonlinear polynomial, a multiobjective vector (F) is defined as below:

$$F = \begin{bmatrix} f_1 \\ f_2 \\ \vdots \\ f_m \end{bmatrix} \quad (5.46)$$

where f_i is defined as Eqn. 5.45, and m is the number of error sensors.

The required noise control system is applied to minimize this objective vector, i.e.

$$\min_{C(z) \in \Phi} F(I) \quad (5.47)$$

where Φ is the searching domain of the controller $C(z)$.

Genetic Active Noise control System. In order to realize Eqn. 5.47, a Genetic Active Noise Control System (GANCS) [217] which is shown in Fig. 5.23 is proposed. GANCS consists of four fundamental units, namely, Acoustic Path Estimation Process (APEP), Genetic Control Design Process (GCDP), Statistic Monitoring Process (SPM) and Decision-Maker (DM).

The design concept of the controller is basically composed of two processes:

1. Acoustic Path Estimation Process for acoustic paths modelling which has the same procedure as Sect. 5.3.1.
2. Genetic Control Design Process for controller design which has a similar structure to that described in Sect. 5.3.2. The main difference in the fitness assignment is explained in more detail in the following sections.

The idea is to model the acoustic paths using FIR filters while the development of the controller relies on this obtained modelling result. The Statistic Monitoring Process monitors the change of environment to ensure the system's robustness as explained in Sect. 5.3.2. The Decision Maker provides an interface so that the problem goal can be defined to fine-tune the optimization process of the GCDP.

Decision Maker. Considering that the Multi-channel Active Noise Control is a multiple objective problem, a Pareto-solution set would be obtained. The Decision Maker selects the solution from this set of non-dominate solutions. In general, the goal attainment method can be applied to achieve the best

5. Genetic Algorithms in Filtering

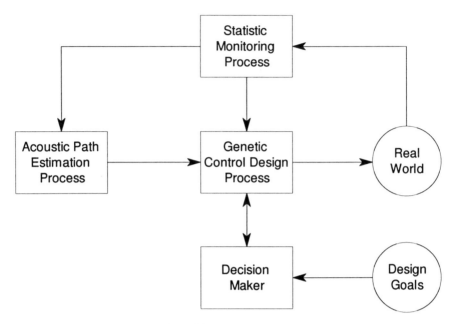

Fig. 5.23. Block diagram of GANCS

global noise suppression result. Consider that the goals of the design objective are expressed as

$$V = (v_1, v_2, \ldots, v_m) \tag{5.48}$$

where v_i is the goal for the design objective f_i.

The non-dominate solutions are compared with the λ values which are expressed below:

$$f_i - \lambda w_i \leq v_i \tag{5.49}$$

where w_i is weighting factor for v_i.

This is the targeted residue noise power level at error sensor i to be reached. Due to different practical requirement, a specific goal should be set. The system will be tuned to meet such goal requirement. This can be illustrated by a simple example:

$$\begin{aligned} w_i &= 1 \quad \forall i \in [1, m] \subset Z \\ v_i &= \alpha \\ v_j &= 2\alpha \end{aligned}$$

which means a higher requirement of silence is assigned in error sensor i.

5.3 Active Noise Control

GCDP – Multiobjective Fitness Assignment. The structure of the GCDP is similar to that of the GA applied in the previous section. The only difference is that a multiobjective rank-based fitness assignment method, as explained in Sect. 3.2, is applied. With the modified ranking method, the non-dominate solution set can be acquired.

Experiment Results. Experiments were carried out to justify the multiobjective approach. A single tone noise signal of 160Hz was used to demonstrate its effectiveness. The error microphones, P1 and P2 were placed 1.3m and 0.5m above ground, respectively. The overall geographic position of the set-up is shown in Fig. 5.24.

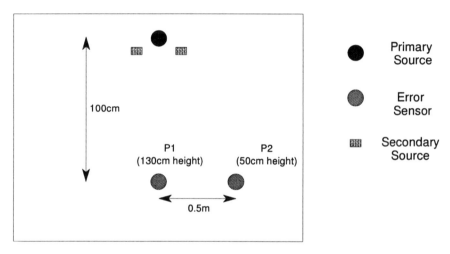

Fig. 5.24. Experimental set-up

The purpose of this particular set-up was to justify the multiobjective approach in ANC, therefore, three tests were conducted:

1. to optimize P1 location only;
2. to optimize P2 location only; and
3. to optimize P1 and P2 simultaneously using the multiobjective approach

Figs. 5.25 and 5.26 depict the noise reduction spectra of P1 and P2 for case (1) and case (2), respectively. Fig. 5.27 shows the compromising effect when both P1 and P2 are optimized by the multiobjective approach. From Table 5.10, it can be observed that the reduction levels on P1 and P2 are under consideration for a simultaneous optimization procedure when the multiobjective approach is applied.

112   5. Genetic Algorithms in Filtering

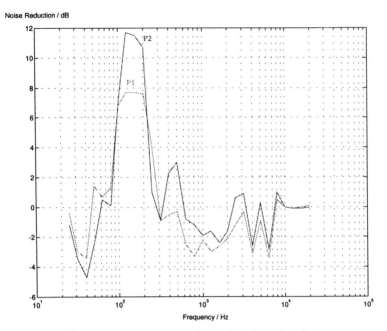

Fig. 5.25. Noise reduction spectra for optimizing P1 only

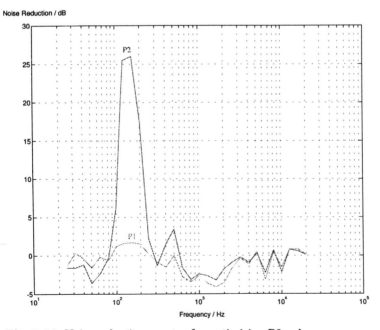

Fig. 5.26. Noise reduction spectra for optimizing P2 only

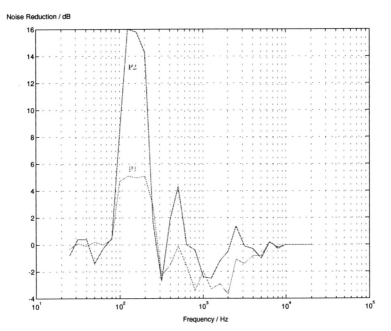

Fig. 5.27. Noise reduction spectra for optimizing P1 and P2

Table 5.10. Noise reduction for different cases

	Position 1		Position 2	
	A_p/dB	$A_p(\omega)/dB$	A_p/dB	$A_p(\omega)/dB$
Noise without Control	92.0	78.8	87.7	74.5
Optimization on P1	84.4 (-7.6)	71.2 (-7.6)	76.7 (-11.0)	63.2 (-11.3)
Optimization on P2	90.5 (-1.5)	77.2 (-1.6)	67.5 (-20.2)	54.9 (-19.6)
MO Approach	87.0 (-5.0)	73.9 (-4.9)	72.7 (-15.0)	59.2 (-15.3)

114   5. Genetic Algorithms in Filtering

To further demonstrate the effectiveness of the multiobjective approach for ANC, the previous experimental set-up was altered so that the new geographical sensor placements were arranged as shown in Fig. 5.28. The particular layout was purposefully arranged in a different form from the one above. The idea was to test the functionality of the multiobjective approach such that the wavefront in this case would be substantially different since the microphone P1 was placed 0.8m ahead of P2. In this way, the noise reduction result would have to be compromised in a more weighted manner than the previous case.

Fig. 5.29 shows the result of noise reduction for optimizing the location at P1 only and Fig. 5.30 depicts the equivalent result at P2. A typical result obtained (candidate 4 in Table 5.11) from the multiobjective approach is shown Fig. 5.31. The overall noise reduction results are tabulated in Table 5.11.

It can be seen from the results obtained in this case that they are quite different when compared to the previous case. There was some noise reduction at P2 while P1 was optimized, but the situation was not reversed at P2 for P1. From Table 5.11, it was a slander for P1 as there was a slight increase of noise level at that point.

However, much more compensated results were obtained from the multiobjective approach. There were five possible solutions which were yielded in the population pool by the effort of the GA. These are also tabulated in Table 5.11. It can be seen that candidate 4 would be an evenly balanced result for P1 and P2 as there is only 1–2 dB difference between them. On the other hand, candidate 1 is considered to be at the other extreme. While P2 reached a 20dB reduction, P1 was considered inactive in the case. The result obtained from candidate 3 is considered to be similar to candidate 4. The candidates 2 and 5 possess the results of a mirror image of each other at P1 and P2. The overall results of the multiobjective approach can be summarized by the trade-off curve as shown in Fig. 5.32. It can be concluded that all five candidates are practically applicable, but the final selection has to be decided by the designer via the decision maker so that the best selected candidate is used to suit that particular environment.

5.3.4 Parallel Genetic Algorithm Approach

While we have tackled the multiobjective issue of the GA in an ANC, our effort is now turned to the computational problem of the GA in an ANC. One of the problems that it is usually encountered with sequential computation of a GA is that it is generally recognized as a slow process. Such a deficiency can be readily rectified by implementing parallel computing architecture. Considering that the GA itself already possesses intrinsic parallelism characteristics,

5.3 Active Noise Control   115

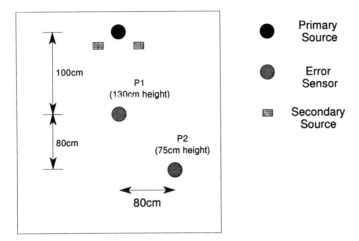

Fig. 5.28. Another experimental set-up

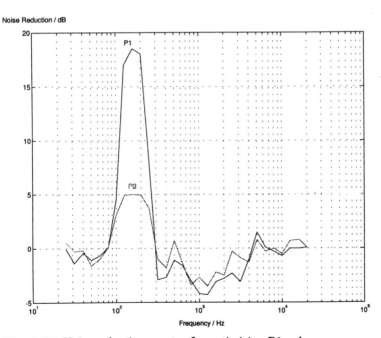

Fig. 5.29. Noise reduction spectra for optimizing P1 only

116   5. Genetic Algorithms in Filtering

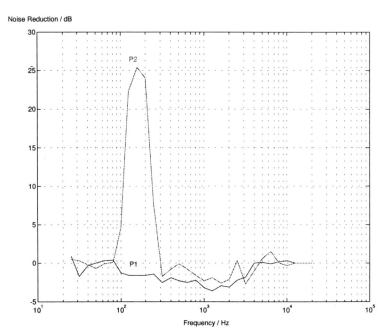

Fig. 5.30. Noise reduction spectra for optimizing P2 only

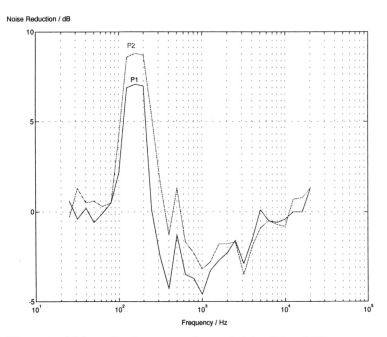

Fig. 5.31. Noise reduction spectra for optimizing P1 and P2

Table 5.11. Noise reduction for different cases

	Position 1		Position 2	
	A_p/dB	$A_p(\omega)/dB$	A_p/dB	$A_p(\omega)/dB$
Noise without Control	91.3	78.0	92.6	79.3
Optimization on P1	73.9 (-17.4)	61.8 (-16.2)	87.7 (-4.9)	74.4 (-4.9)
Optimization on P2	92.9 (+1.6)	79.6 (+1.6)	71.3 (-21.3)	58.1 (-21.2)
MO Candidate 1	91.4 (+0.1)	78.1 (+0.1)	72.0 (-20.6)	58.8 (-20.5)
MO Candidate 2	88.7 (-2.6)	75.4 (-2.6)	79.9 (-12.7)	66.6 (-12.7)
MO Candidate 3	85.7 (-5.6)	72.6 (-5.4)	82.0 (-10.6)	68.6 (-10.7)
MO Candidate 4	84.2 (-7.1)	71.1 (-6.9)	83.9 (-8.7)	70.6 (-8.7)
MO Candidate 5	73.8 (-17.5)	61.9 (-16.1)	87.1 (-5.5)	73.8 (-5.5)

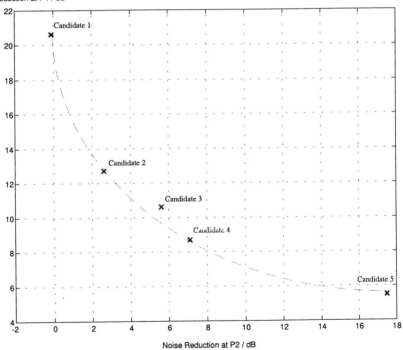

Fig. 5.32. Noise reduction trade-off between P1 and P2 due to multiobjective approach

it requires no extra effort to construct such parallel architecture to gain the required speed for practical uses. There are a number of GA-based parallel methods that can be used to enhance computational speed (as explained in Sect. 3.1). The global parallel GA (GPGA) is recommended for this application, in the sense that a multiple of system response is guaranteed.

The problem of a slow GA computation is not really caused by genetic operations such as selection, recombination, mutation, fitness assignments and so on. In fact, any one of these operations requires little computational time. The main contributor to the time consumption is the actual numeric calculation of the objective functions. This is particularly apparent when a number of objective functions have to be simultaneously optimized such that their functions may even be nonlinear, constrained and discontinuous.

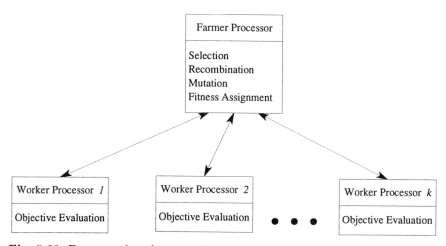

Fig. 5.33. Farmer-and-workers structure

Fig. 5.33 shows the farmer-and-worker structure for such global parallelism architecture which is well suited for this application.

The GPGA treats the entire population as a single breeding mechanism. The purpose of the farmer-processor (FP), being a master control unit, is to generate the new chromosomes while the worker-processors (WP) are used to evaluate the objective values of the chromosomes. The objective values of the new chromosomes are then returned to the FP for the process of reinsertion and the necessary fitness assignment. In this way, the time-demanding process of objective evaluation is now handled by a dedicated WP, and hence improves the computational speed in the order of a multiple fashion if a number of WPs are used.

Parallel Hardware Approach. Based on the farmer-and-worker parallel structure, a scalable multiple digital signal processing (DSP) based parallel hardware architecture has been developed [103]. The overall parallel architecture is shown in Fig. 5.34.

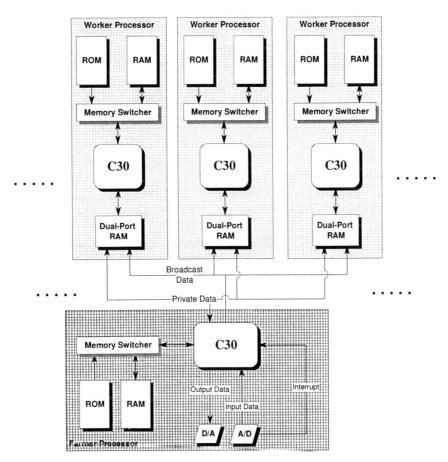

Fig. 5.34. Active noise control system

This architecture was formed by the use of Texas Instruments' TMS320C30 (C30) DSP processors. A C30 processor was designed as a FP unit and this handles all the input/output (IO) activities, as well as generating the new chromosomes. The FP also co-ordinates the data flow between the real-world and the system. By the use of the C30's two independent external bus interface ports: Primary Bus and Expansion Bus, the external IO address, local memory address and WP address are separated. The operations of the two buses are basically identical except that the Primary Bus address space that is 24-bit in width and the Expansion Bus address space that is 13-bit

120   5. Genetic Algorithms in Filtering

in width. The local memory and all the WPs can be accessed through the Primary Bus. However, all the IO peripherals are interfaced by the Expansion Bus.

The objective values of the chromosomes are evaluated by the C30s as WPs. These values return to the FP for the process of reinsertion and fitness assignment. Each WP has its own local memory units and associated dual-port RAM. The reason for using this device is to reduce multi-DSP interfacing complexity and to provide a scalable feature so that the inter-processor communication bandwidth can thus be increased. The other advantage of using a dual-port RAM is that it ensures the FP can communicate with the WP in the conventional memory access operating manner.

In this configuration, the main action of the FP is only to read/write data on the different memory segments so that the information can be easily accessed by a number of WPs. In our design, the memory size of the dual-port RAM is 2K words and the address Bus width of the FP is 24 bits, which implies that 16M bytes memory can be accessed through the Primary Bus. Therefore, the maximum number of WPs that can be handled by the FP is 8192 (16M/2K)!

As the number of WPs increases, so the interprocessor communication overhead increases substantially. In order to reduce the communication overhead, two types of data are classified: broadcast data and private data. Each type of data uses its own addressing method. The broadcast data are used typically during the initialization stage when all the initial data and parameters are fed into WPs. Therefore, as the number of WPs increases, the initialization processing time remains constant. The private data type is used by the FP to update the individual WP's optimized result.

This GA hardware platform using DSP chips should serve as an ideal test-bed facility for system design and development before dedicated system-specific hardware is built.

Filter Model. Another filter model, IIR filter, was tried in this experiment. Due to the multimodality of the IIR filter error surface, the GA is well suited to optimize the filter coefficients to search for global optima [157, 233]. An IIR filter can be constructed in lattice form and this structure is shown in Fig. 5.35.

Similarly, the delay and gain elements are already embedded in this filter model

$$H(z) = gz^{-m}\frac{B(z)}{A(z)} \qquad (5.50)$$

5.3 Active Noise Control

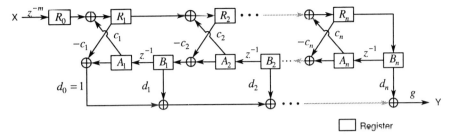

Fig. 5.35. Lattice form of IIR filter

$$= gz^{-m} \frac{\sum_{i=0}^{n} b_i z^{-i}}{\sum_{j=0}^{n} a_j z^{-j}} \qquad (5.51)$$

where m is an additional delay; g is the gain; a_i, b_i are the IIR coefficients ($a_0 = b_0 = 1$), and c_i, d_i are the lattice filter coefficients which can be obtained by

$$c_i = a_i^{(i)}$$

$$d_i = b_i + \sum_{j=i+1}^{n} d_j a_{j-i}^{(j)}$$

with $a_i^{(n)} = a_i$, $\forall i = 1, 2, \ldots n$ and the recursive function

$$a_j^{(i-1)} = \frac{a_j^{(i)} + c_i a_{i-j}^{(i)}}{1 - c_i^2} \qquad (5.52)$$

Referring to Fig. 5.35, for a n-poles and n-zeros system, we can obtain the following equations.

$$R_0 = x(k-m) \qquad (5.53)$$
$$R_i = R_{i-1} + C_i A_i \quad i = 1, \ldots, n \qquad (5.54)$$
$$B_i = -C_{i+1} R_{i+1} + A_{i+1} \quad i = 1, \ldots, n-1 \qquad (5.55)$$
$$B_n = R_n \qquad (5.56)$$

The output of this lattice IIR filter is computed by Eqn. 5.57.

$$y(k) = g \left\{ -c_1 R_1 + A_1 + \sum_{i=1}^{n} d_i B_i \right\} \qquad (5.57)$$

The principle advantage of lattice filters over alternative realizations such as Direct-Form, parallel and cascade, is that stability can be maintained simply by restricting the lattice reflection coefficients to lie within the range +1 to -1. In addition, the lattice form is known to be less sensitive to coefficient round-off. There are totally $(2n+2)$ parameters (m, g), $(c_1, c_2, \ldots, c_n)$, $(d_0, d_1, \ldots, d_{n-1})$ to optimize for a n-th order IIR filter.

Chromosome Representation. This is used to optimize the IIR parameters using a GA. These parameters can be coded into genes of the chromosome for GA optimization. There are two different data types. The unsigned integer delay gene, $[m]$, is represented by a 16-bit binary string and each parameter in the real-valued coefficient genes $[g, c_i, d_i]$ is represented by a 32-bit binary string. The parameters are constrained as

$$\begin{aligned} m &\in [0, m_{max}] \subset Z^+ \\ |g| &\in [g_{min}, g_{max}] \subset \Re^+ \\ C &= [c_1, c_2, \ldots, c_n] \in (-1, 1)^n \subset \Re^n \\ D &= [d_1, d_2, \ldots, d_n] \subset \Re^n \end{aligned} \quad (5.58)$$

where $m_{max} = 100$, $g_{min} = 0.01$, and $g_{max} = 2.0$.

The structure of the chromosome (I) is thus formulated as follows:

$$I = \{m, g, C, D\} \in \Phi \subset Z \times \Re \times \Re^n \times \Re^n \quad (5.59)$$

where Φ is the searching domain of the controller $C(z)$.

For this application, 4th-order IIR filters were used to model the acoustic path and the controller. Hence, $n = 4$. With similar operations, the GA is ready to go.

Experimental Results. This formulation was used for the investigation. The ANC noise reduction result due to this approach is shown in Fig. 5.36. It can be demonstrated that the parallelism of the system could greatly reduce the computation time with a number of worker processors.

5.3.5 Hardware GA Processor

Thus far, the study of the GA in an ANC application has been successfully carried out. The results are very encouraging for developing an industrial/commercial product. However, the cost for realizing the goal is currently quite expensive, as the computers or signal processors etc. that have been incurred in the study would not be economically viable. Furthermore, for an industrial product, the ANC system must be implemented in a VLSI chip version in order to be marketable. Therefore, some means of cost cutting exercise must be sought in order to translate this technology into practice.

However, the development cost of VLSI chip does not present less cost in a dollar sense. The number of iterations of development must be limited to a minimum to avoid budgeting over expense. Therefore, the development of a low cost version by the use of Field Programmable Gate Arrays (FPGA) could be ideally applied. The version of development can be realized based on the

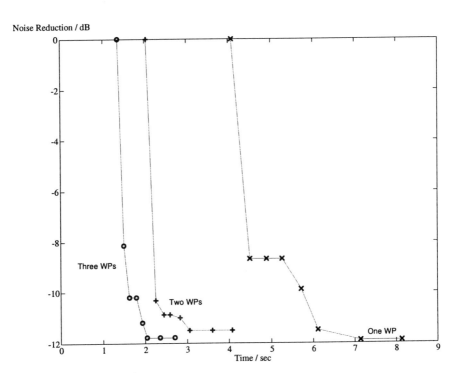

Fig. 5.36. System performance

methodology that has already been described. This work is now being actively carried out, and the implementation outlined here illustrates its feasibility for the development of the future ANC systems.

It has become clear that the GA provides a means to evolve solutions to search and optimize problems instead of using other fixed mathematically derived algorithmic procedures. It is also well understood that the GA possesses a number of computational dependable schemes in order to fulfil its goal. To further improve the performance of the GA in terms of computation time, a hardware GA processor is therefore a clear solution. One of the best ways to implement a GA in hardware is to construct a model using the hardware descriptive language VHDL and to synthesize the model in FPGA technology. The realization of Global GA, see Fig. 3.1, and Migration GA, see Figs. 3.2–3.4, can be easily achieved. In this particular design, the Actel FPGA [1] using Synopsys [183] compiler has been adopted. Fig. 5.37 shows the basic design flow of synthesizing an VHDL design description into an FPGA using the Synopsys Design Compiler, FPGA Compiler, or Design Analyzer.

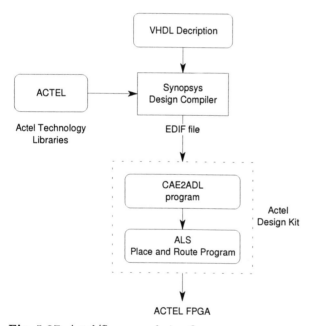

Fig. 5.37. Actel/Synopsys design flow

This design flow is based on the Parallel Global GA that is described in Sect. 5.3.4. This architecture is particularly useful for Single-Instruction-

Multiple-Data (SIMD) design in which all the offspring are able to proceed with the same fitness evaluation process. The overall hardware for the Hardware GA processor is shown in Fig. 5.38. This is a flexible modular structure that can be further expanded if necessary. This architecture consists of three different modules, namely Fitness Evaluator (FE), Objective Function Sequencer (OFS) and Genetic Operators (GO). Each of the modules is implemented with an FPGA. This modular structure can easily lead to varied application environments by replacing a suitable module, and yet, the demand of computation power is met.

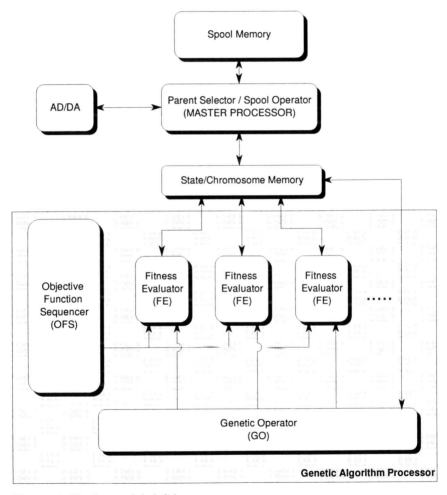

Fig. 5.38. Hardware global GA

126   5. Genetic Algorithms in Filtering

The relationship between the operation and inter-module of the GAP can be described as follows: In each GA cycle, the parents are selected by the external processing unit (parent selector). These parents are then fed to the GO module. A simplified structure diagram for the GO is shown in Fig. 5.39. Two GA operations, namely the mutation and crossover, are processed by the GO while a uniform crossover is adopted to exchange the genetic information of the parents. The action of mutation is then applied for altering the genes randomly with a small probabilistic value. Knowing that one difficulty of the GO would be the hardware implementation of a realistic Gaussian distributed random number generator in the FPGA. A linear distributed function is therefore proposed for such a design. The details of this implementation are described in Appendix C.

Once the crossover and mutation operations have been completed in the GO, the required offspring are born. The quality of these offspring is then examined by the FE via a fitness evaluation process. This is usually a cost function in the least square form. In this case, the hardware construction involves only the development of adders and multipliers which can be easily constructed by the use of FPGA chips. The actual sequence of calculating the fitness values is specified by the instructions of a sequencer design of the OFS.

The architecture of the OFS can also be built using the FPGA chip. The OFS provides the sequence to perform the fitness evaluation of the FE. This evaluation is only completed when the fitness value of each offspring has been calculated. At this stage, all the offspring, together with their fitness values are then fed into the external state/chromosome memory for insertion into the population pool.

This loosely coupled module structure enables the GAP design to be easily fitted into various application environments in order to suit different types of computation power requirements. Since this architecture has been purposely designed not only for ANC, but also for general engineering use, the design of the OFS sequencer can be changed according to the style of the fitness functions, which in itself is a problem dependent element.

The advantage of this set up is that the units of the FE and the GO remain unchanged, despite the fact that they may be used for other applications. Furthermore, it is a scalable architecture which provides enhanced versatility for the GAP to be designed to tackle complex real-time applications. As for the time-demand problem, more FEs can be added in order to fulfil the computation requirement.

For the purposes of Active Noise Control (ANC), the lattice filter model is adopted, and the formulation of the chromosome is based on the Sect. 5.3.4.

5.3 Active Noise Control 127

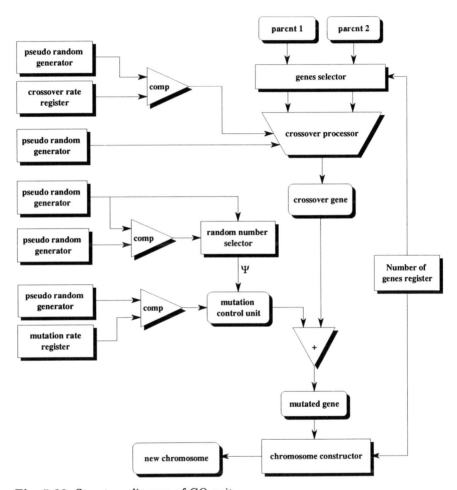

Fig. 5.39. Structure diagram of GO unit

Fitness Function. The fitness function reflects the likeness between the output of the estimated model and the real output. In mathematical terms, it is expressed as:

$$f = \sum_{i=1}^{N}(y_i - \hat{y}_i)^2 \tag{5.60}$$

where $\hat{y}_i$ is the estimated output from a filter model; y_i is the real output; and N is the sample window. In our experiments, $N = 400$.

Fitness Evaluator. To realize Eqn. 5.60, the design of an FE involves the design of a multiplier and an adder. A development of a high speed multiplier is the major effort for improving the computing speed. Hence, it is the main core of the FE design. In our design, a high pipelined multiplier is proposed as indicated in Fig. 5.40. The multiplier algorithm is based on the redundant binary representation [213], which is described in Appendix D. To compensate for the propagation delay of the logic modules in the FPGA, a three-stage pipeline is applied for the design to improve the performance speed.

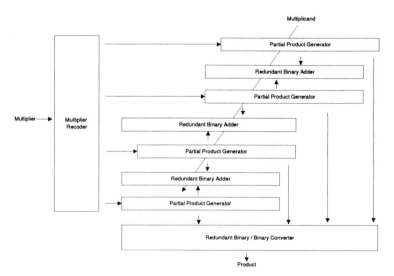

Fig. 5.40. Block diagram of the multiplier

Objective Function Sequencer. Based on the pipeline structure, a good sequencer design means a high efficiency of multiplier pipeline usage. As a consequence, this improves the GAP's overall performance. Therefore, it was necessary to compromise between the system performance (filter order) and the computation demand (number of steps required for completing the evaluation).

5.3 Active Noise Control

In our design, a 4x4 IIR lattice filter based on Eqn. 5.57 was applied. A computation step table, Table 5.12, was formed. The number of steps needed to complete one cycle of the fitness function evaluation are stated in this table. There are 15 computation steps in total. Since a three-stage pipeline multiplier is designed in the FE, the executing sequences are stated in Table 5.13 which was developed on the basis of Table 5.12.

Table 5.12. The computation step table of the 4X4 IIR lattice filter

state0	read x
state1	tmp1=x+c1a1
state2	tmp=tmp1+c2a2
state3	add out=a1g
state4	tmp1=add out-c1gtmp1
state5	tmp1=tmp+c3a3
state6	tmp=tmp1+c4a4
state7	a1=a2-c2tmp
state8	a2=a3-c3tmp1
state9	a3=a4-c4tmp
state10	add out=tmp1+d4tmp, read y
state11	add out=add out-y
state12	add out=add out+d1a1
state13	add out=add out+d2a2, a4=tmp
state14	Acc IN=add out+d3a3
state15	read x, goto state1

The elementary multiplication and addition operations, including the registers used for each step in Table 5.12 were firstly identified. Since this basic computation operation and register usage information had already been extracted, the register/output usage dependency was therefore known. According to this dependency information and the number of pipelines used in the multiplier, the execution sequence of the multiplier and adder had to be rearranged so that the pipeline of the multiplier was fully utilized. As a result, the throughput of the FE was thus maximized. The final FE internal execution table is tabulated in Table 5.13.

Genetic Operators. The genetic operation settings required to perform the selection, crossover, mutation, population size etc. are tabulated in Table 5.14.

System Performance. The speed of computation by FE is largely dependent upon the number of FEs being used. An independent study was carried out to investigate the relationship between the time taken to complete the number of iterations with the various numbers of FEs used. The result is summarized in Fig. 5.41. This undoubtedly demonstrates that the parallel architecture of the FE can greatly reduce computation time.

Table 5.13. The FE internal execution sequence table of the 4X4 IIR lattice filter

Mul A	Mul B	Mul P	Add In	Add Out	Add Op	Transfer
C1	TMP					
D1	TMP					
D2	TMP					
D3	TMP	C1G	'0'		Add	
D4	TMP	D1G	'0'	C1G	Add	
C1	A1	D2G	'0'	D1	Add	
C2	A2	D3G	'0'	D2	Add	
C3	A3	D4G	'0'	D3	Add	READ X
C4	A4	C1A1	X	D4	Add	
G	A1	C2A2	Add Out	TMP1	Add	
C1G	TMP1	C3A3	Add Out	TMP	Add	
C2	TMP	C4A4	Add Out	TMP1	Add	
C3	TMP1	A1G	'0'	TMP	Add	
C4	TMP	C1GTMP1	Add Out		Sub	
D4	TMP	C2TMP	A2	TMP1	Sub	READ Y
Y	'1'	C3TMP1	A3	A1	Sub	
D1	A1	C4TMP	A4	A2	Sub	
D2	A2	D4TMP	TMP1	A3	Add	A4=TMP
D3	A3	Y	Add Out		Sub	
C1	A1	D1A1	Add Out		Add	
C2	A2	D2GA2	Add Out		Add	
C3	A3	D3GA3	Add Out		Add	READ X
C4	A4	C1A1	X	Acc IN	Add	
.	.	C2A2	Add Out	TMP1	Add	
.	.	C3A3	Add Out	TMP	Add	
.	.	C4A4	Add Out	TMP1	Add	
.	.	.	.	TMP	.	

Table 5.14. Genetic settings for ANC system

Population Size	30
Offspring generated per cycle	4
Selection	ranking
Crossover	Uniform crossover (rate=0.9)
Mutation	random mutation (rate = 0.1)

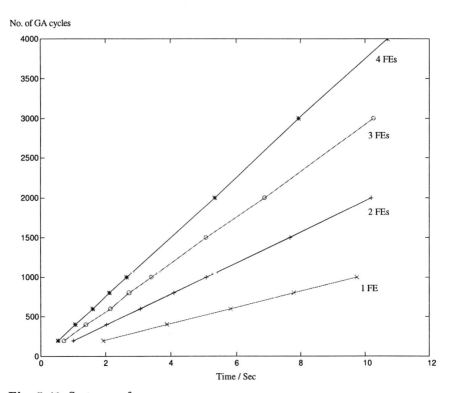

Fig. 5.41. System performance

6. Genetic Algorithms in H-infinity Control

During the last decade or so, $\mathbf{H}^\infty$ optimization has emerged as a powerful tool for robust control system design. This has a sound theoretical background for handling model uncertainties. Based on $\mathbf{H}^\infty$ optimization, a variety of design methods can be developed. The $\mathbf{H}^\infty$ LSDP is one of that has proven to be effective in practical industrial design. The approach involves the robust stabilization to the additive perturbations of normalized coprime factors of a weighted plant. Prior to robust stabilization, the open-loop singular values are also shaped using weighting functions to give a desired open-loop shape which corresponds to good closed-loop performance. However, a successful design using LSDP depends on the appropriate choice of weighting functions, which in turn relies on a designer's experience and familiarity with the design approach.

In [230], it is proposed to enhance the LSDP by combining it with numerical optimization techniques. In order to more effectively search for optimal solutions to the derived constrained optimization problems, the Multiple Objective Genetic Algorithm is suggested in [231]. In this mixed optimization approach, the structures of the weighting functions are to be pre-defined by the designer. It is not possible to search an optimal design systematically among the various structured weights. Therefore, the HGA is a perfect approach to address such a problem. In this chapter, two distinguished design case studies, i.e. distillation column design [134] and the universal benchmark process [85] have been chosen for the demonstration of HGA effectiveness in H-infinity control.

6.1 A Mixed Optimization Design Approach

LSDP is based on the configuration as depicted in Fig. 6.1, where $(\tilde{N}, \tilde{M}) \in R\mathbf{H}^\infty$, the space of stable transfer function matrices, is a normalized left coprime factorization of the nominal plant G. That is, $G = \tilde{M}^{-1}\tilde{N}$, and $\exists V, U \in R\mathbf{H}^\infty$ such that $\tilde{M}V + \tilde{N}U = I$, and $\tilde{M}\tilde{M}^* + \tilde{N}\tilde{N}^* = I$; where for a real rational function of s, X^* denotes $X^T(-s)$.

6. Genetic Algorithms in H-infinity Control

For a minimal realization of $G(s)$

$$G(s) = D + C(sI - A)^{-1}B \stackrel{s}{=} \left[\begin{array}{c|c} A & B \\ \hline C & D \end{array}\right] \tag{6.1}$$

a normalized coprime factorization of G can be given by [147]

$$[\tilde{N} \ \tilde{M}] \stackrel{s}{=} \left[\begin{array}{c|cc} A + HC & B + HD & H \\ \hline R^{-1/2}C & R^{-1/2}D & R^{-1/2} \end{array}\right] \tag{6.2}$$

where $H = -(BD^T + ZC^T)R^{-1}$, $R = I + DD^T$, and the matrix $Z \geq 0$ is the unique stabilizing solution to the algebraic Riccati equation (ARE)

$$(A - BS^{-1}D^TC)Z + Z(A - BS^{-1}D^TC)^T$$
$$- ZC^TR^{-1}CZ + BS^{-1}B^T = 0 \tag{6.3}$$

where $S = I + D^TD$

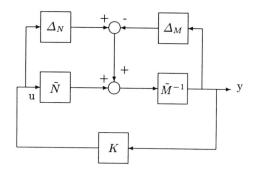

Fig. 6.1. Robust stabilization with respect to coprime factor uncertainty

A perturbed model G_p of G is defined as

$$G_p = (\tilde{M} + \Delta_M)^{-1}(\tilde{N} + \Delta_N) \tag{6.4}$$

where $\Delta_M, \Delta_N \in R\mathbf{H}^\infty$.

To maximize the class of perturbed models such that the closed-loop system in Fig. 6.1 is stabilized by a controller $K(s)$, the $K(s)$ must stabilize the nominal plant G and minimize γ [76] where

$$\gamma = \left\| \begin{bmatrix} K \\ I \end{bmatrix} (I - GK)^{-1} \tilde{M}^{-1} \right\|_\infty \tag{6.5}$$

From the small gain theorem, the closed-loop system will remain stable if

$$\| [\Delta_N \ \Delta_M] \|_\infty < \gamma^{-1} \tag{6.6}$$

6.1 A Mixed Optimization Design Approach

The minimum value of γ, (γ_0), for all stabilizing controllers is given by

$$\gamma_0 = (1 + \lambda_{max}(ZX))^{1/2} \tag{6.7}$$

where $\lambda_{max}(\cdot)$ represents the maximum eigenvalue, and $X \geq 0$ is the unique stabilizing solution to the following ARE

$$(A - BS^{-1}D^T C)^T X + X(A - BS^{-1}D^T C) \\ - XBS^{-1}B^T X + C^T R^{-1}C = 0 \tag{6.8}$$

A controller which achieves a γ is given in [147] by

$$K \stackrel{s}{=} \left[\begin{array}{c|c} A + BF + \gamma^2(Q^T)^{-1}ZC^T(C+DF) & \gamma^2(Q^T)^{-1}ZC^T \\ \hline B^T X & -D^T \end{array} \right] \tag{6.9}$$

where $F = -S^{-1}(D^T C + B^T X)$ and $Q = (1 - \gamma^2)I + XZ$.

A descriptor system approach may be used to synthesize an optimal controller such that the minimum value γ_0 is achieved.

In practical designs, the plant needs to be weighted to meet closed-loop performance requirements. A design method, known as the LSDP, has been developed [147, 148] to choose the weights by studying the open-loop singular values of the plant, and augmenting the plant with weights so that the weighted plant has an open-loop shape which will give good closed-loop performance.

This loop shaping can be achieved by the following design procedure:

1. Using a pre-compensator, W_1, and/or a post-compensator, W_2, the singular values of the nominal system G are modified to give a desired loop shape. The nominal system and weighting functions W_1 and W_2 are combined to form the shaped system, G_s, where $G_s = W_2 G W_1$. It is assumed that W_1 and W_2 are such that G_s contains no hidden unstable modes.
2. A feedback controller, K_s, is synthesized which robustly stabilizes the normalized left coprime factorization of G_s, with a stability margin ϵ, and
3. The final feedback controller, K, is then constructed by combining the H^∞ controller K_s, with the weighting function W_1 and W_2 such that

$$K = W_1 K_s W_2.$$

For a tracking problem, the reference signal is generally fed between K_s and W_1, so that the closed loop transfer function between the reference r and the plant output y becomes

$$Y(s) = (I - G(s)K(s))^{-1}G(s)W_1(s)K_s(0)W_2(0)R(s), \tag{6.10}$$

with the reference r is connected through a gain $K_s(0)W_2(0)$ where

$$K_s(0)W_2(0) = \lim_{s \to 0} K_s(s)W_2(s), \tag{6.11}$$

to ensure unity steady state gain.

The above design procedure can be developed further into a two-degree-of-freedom (2 DOF) scheme as shown in Fig. 6.2.

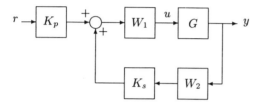

Fig. 6.2. The 2 DOF scheme

The philosophy of the 2 DOF scheme is to use the feedback controller $K_s(s)$ to meet the requirements of internal stability, disturbance rejection, measurement noise attenuation, and sensitivity minimization. The pre-compensator K_p is then applied to the reference signal, which optimizes the response of the overall system to the command input. The pre-compensator K_p depends on design objectives and can be synthesized together with the feedback controller in a single step via the $\mathbf{H}^\infty$ LSDP [110].

In LSDP, the designer has the freedom to choose the weighting functions. Controllers are synthesized directly. The appropriate weighting functions will generate adequate optimal γ_o and will produce a closed-loop system that is robust and offers satisfactory and non-conservative performance. The selection of weighting functions is usually performed by a trial-and-error method and is based on the designer's experience. In [230], it is proposed that the Method of Inequalities (MOI) [246] be incorporated with LSDP, such that it is possible to search for "optimal" weighting functions automatically to meet more explicit design specifications in both the frequency-domain and the time-domain.

In this mixed optimization approach, the weighting functions W_1 and W_2 are the *design parameters*. Control system design specifications are given in a set of inequality constraints. That is, for a given plant $G(s)$, to find (W_1, W_2) such that

$$\gamma_o(G, W_1, W_2) = (1 + \lambda_{max}(ZX))^{1/2} \leq \varepsilon_\gamma \tag{6.12}$$

and
$$\phi_i(G, W_1, W_2) \leq \varepsilon_i; \quad i = 1, 2, \ldots, n \tag{6.13}$$

where ϕ_i's are performance indices, which are algebraic or functional inequalities representing rise-time, overshoot, etc. and ε_γ and ε_i are real numbers representing the desired bounds on γ_0 and ϕ_i, respectively.

Numerical search algorithms may then be applied to find solutions to the above optimization problems.

6.1.1 Hierarchical Genetic Algorithm

The constrained optimization problems derived by the mixed optimization design approach described in Sect. 6.1 are usually non-convex, non-smooth and multiobjective with several conflicting design aims which need to be achieved simultaneously. In [231], a Multiobjective Genetic Algorithm [63] is employed to find solutions to such optimization problems. Successful designs have been achieved. It has, however, been found that due to the order and structure of weighting functions being pre-fixed by the designer, the approach lacks flexibility in choosing weights of different forms, which in turn affects the optimality of the design. The HGA has thus been considered to optimize both the orders and coefficients of the weights, W_1 and W_2, used in the design.

Objective Function and Fitness Function. An objective function f is defined, for a chromosome if Eqn. 6.12 is satisfied, as the number of violated inequalities in Eqn. 6.13. The procedure of objective function evaluation is listed as follows:

1. For a chromosome $I = (W_1, W_2)$ in hierarchy coded form, generate the corresponding W_1 and W_2
2. Calculate $G_s = W_2 G W_1$
3. Find the solutions Z_s, X_s to Eqns. 6.3 and 6.8
4. Calculate $\gamma_0(W_1, W_2)$ by Eqn. 6.12
5. Synthesize K_s by Eqn. 6.9
6. Calculate $\phi_i(G, W_1, W_2)$ for the present chromosome; and
7. Compute f by

$$f = \begin{cases} \sum_{i=1}^n m_i & \text{if } \gamma_0 < \varepsilon_\gamma \\ n + 1 + \gamma_0 & \text{else} \end{cases} \tag{6.14}$$

where
$$m_i = \begin{cases} 0 & \text{if } \phi_i \leq \varepsilon_i \\ 1 & \text{else} \end{cases} \tag{6.15}$$

To convert the objective function (f) to the fitness value, a linear ranking approach [234] is applied.

Genetic Operations. Crossover and Mutation on the binary string [47] are applied independently to different levels of a chromosome as in a standard GA.

Optimization Procedure. The optimization procedure is listed as follows:

1. Define the plant G and define the functions ϕ_i;
2. Define the values of ε_i and ε_γ;
3. Define the fundamental form of the weighting functions W_1 and W_2, and the search domain of R_1, R_2;
4. Define the parameters for the HGA;
5. Randomly generate the first population;
6. Calculate the objective value and assign the fitness value to each chromosome;
7. Start the HGA cycle
   - Select parents by the Stochastic Universal Sampling method [10],
   - Generate new chromosomes via Crossover and Mutation,
   - Calculate the objective values of the new chromosomes,
   - Reinsert the new chromosomes into the population and discard the same number of old, low-ranked chromosomes;
8. Terminate if Eqns. 6.12 and 6.13 are satisfied, otherwise repeat the HGA cycle.

6.1.2 Application I: The Distillation Column Design

This proposed algorithm has been used to design a feedback control system for the high-purity distillation column described in [134]. The column was considered in its LV* configuration [202], for which the following model was relevant

$$G_D(s, k_1, k_2, \tau_1, \tau_2) = \frac{1}{75s+1} \begin{bmatrix} 0.878 & -0.864 \\ 1.082 & -1.096 \end{bmatrix} \times$$
$$\begin{bmatrix} k_1 e^{-\tau_1 s} & 0 \\ 0 & k_2 e^{-\tau_2 s} \end{bmatrix} \quad (6.16)$$

where $0.8 \leq k_1, k_2 \leq 1.2$ $0 \leq \tau_1, \tau_2 \leq 1$, and all time units were in minutes.

The time-delay and actuator-gain values used in the nominal model G_n were $k_1 = k_2 = 1$ and $\tau_1 = \tau_2 = 0.5$. The time-delay element was approximated by a first-order Padé approximation for the nominal plant. The design specifications are to devise a controller which guarantees for all $0.8 \leq k_1, k_2 \leq 1.2$ and $0 \leq \tau_1, \tau_2 \leq 1$:

1. Closed-loop stability

---

* LV indicates that the inputs used are reflux(L) and boilup (V).

6.1 A Mixed Optimization Design Approach

2. The output response to a step demand $h(t)\begin{bmatrix} 1 \\ 0 \end{bmatrix}$ satisfies $-0.1 \leq y_1(t) \leq 1.1$ for all t, $y_1(t) \geq 0.9$ for all $t > 30$ and $-0.1 \leq y_2(t) \leq 0.5$ for all t.

3. The output response to a step demand $h(t)\begin{bmatrix} 0.4 \\ 0.6 \end{bmatrix}$ satisfies $y_1(t) \leq 0.5$ for all t, $y_1(t) \geq 0.35$ for all $t > 30$ and $y_2(t) \leq 0.7$ for all t, and $y_2(t) \geq 0.55$ for all $t > 30$.

4. The output response to a step demand $h(t)\begin{bmatrix} 0 \\ 1 \end{bmatrix}$ satisfies $-0.1 \leq y_1(t) \leq 0.5$ for all t, $-0.1 \leq y_2(t) \leq 1.1$ for all t and $y_2(t) \geq 0.9$ for all $t > 30$.

5. The frequency response of the closed-loop transfer function between demand input and plant input is gain limited to 50dB and the unity gain crossover frequency of its largest singular value should be less than 150 rad/min.

A set of closed-loop performance functionals $\{\phi_i(G_D, W_1, W_2), i=1,2,\ldots, 16\}$, is then defined accounting for the design specifications given above. Functionals ϕ_1 to ϕ_{14} are measures of the step response specifications. Functionals ϕ_1, ϕ_6, ϕ_8 and ϕ_{11} are measures of the overshoot; ϕ_4, ϕ_5, ϕ_{13} and ϕ_{14} are measures of the undershoot; ϕ_2, ϕ_7, ϕ_9 and ϕ_{12} are measures of the rise-time; and ϕ_3 and ϕ_{10} are measures of the cross-coupling. Denoting the output response of the closed-loop system with a plant G_D at a time t to a reference step demand $h(t)\begin{bmatrix} h_1 \\ h_2 \end{bmatrix}$ by $y_i([\ h_1\ h_2\ ]^T)$, $i = 1,2$ the step-response functionals are

$$\phi_1 = \max_t y_1([\ 1\ \ 0\ ]^T, t) \qquad (6.17)$$

$$\phi_2 = -\min_{t>30} y_1([\ 1\ \ 0\ ]^T, t), \qquad (6.18)$$

$$\phi_3 = \max_t y_2([\ 1\ \ 0\ ]^T, t), \qquad (6.19)$$

$$\phi_4 = -\min_t y_1([\ 1\ \ 0\ ]^T, t), \qquad (6.20)$$

$$\phi_5 = -\min_t y_2([[\ 1\ \ 0\ ]^T, t), \qquad (6.21)$$

$$\phi_6 = \max_t y_1([\ 0.4\ \ 0.6\ ]^T, t), \qquad (6.22)$$

$$\phi_7 = -\min_{t>30} y_1([\ 0.4\ \ 0.6\ ]^T, t), \qquad (6.23)$$

$$\phi_8 = \max_t y_2([\ 0.4\ \ 0.6\ ]^T, t), \qquad (6.24)$$

$$\phi_9 = -\min_{t>30} y_2([\ 0.4\ \ 0.6\ ]^T, t), \qquad (6.25)$$

$$\phi_{10} = \max_t y_1([\ 0\ \ 1\ ]^T, t), \qquad (6.26)$$

$$\phi_{11} = \max_t y_2([\ 0\ \ 1\ ]^T, t), \qquad (6.27)$$

$$\phi_{12} = -\min_{t>30} y_2([\ 0\ \ 1\ ]^T, t) \qquad (6.28)$$

6. Genetic Algorithms in H-infinity Control

$$\phi_{13} = -\min_t y_1([\ 0\ \ 1\ ]^T, t), \qquad (6.29)$$

$$\phi_{14} = -\min_t y_2([\ 0\ \ 1\ ]^T, t) \qquad (6.30)$$

The steady-state specifications are satisfied automatically by the use of integral action. From the gain requirement in the design specifications, ϕ_{15} is the $\mathbf{H}^\infty$-norm (in DB) of the closed-loop transfer function between the reference and the plant input.

$$\phi_{15} = \sup_\omega \bar\sigma\left((I - K(j\omega)G_D(j\omega))^{-1} W_1(j\omega)K_s(0)W_2(0)\right) \qquad (6.31)$$

From the bandwidth requirement in the design specification, ϕ_{16} is defined (in rad/min) as

$$\phi_{16} = \max\{\omega\} \text{ such that}$$
$$\bar\sigma\left((I - K(j\omega)G_D(j\omega))^{-1} W_1(j\omega)K_s(0)W_2(0)\right) \geq 1 \qquad (6.32)$$

The fundamental structures of W_1 and W_2 in the design examples are given as:

$$W_1 = \frac{(s+w_5)(s+w_6)(s^2+w_7s+w_8)}{s(s+w_1)(s+w_2)(s^2+w_3s+w_4)}\begin{bmatrix}\alpha_1 & 0\\ 0 & \alpha_2\end{bmatrix} \qquad (6.33)$$

$$W_2 = \frac{(s+w_{13})(s+w_{14})(s^2+w_{15}s+w_{16})}{(s+w_9)(s+w_{10})(s^2+w_{11}s+w_{12})}\begin{bmatrix}\alpha_3 & 0\\ 0 & \alpha_4\end{bmatrix} \qquad (6.34)$$

In general, W_1 and W_2 can be diagonal matrices with different diagonal elements.

The chromosome is a binary string describing the control and coefficient genes, g_c and g_r where

$$g_c \in B^{12}$$
$$g_r = \{w_1, w_2, \ldots, w_{16}, \alpha_1, \alpha_2, \alpha_3, \alpha_4\} \in R_1^{16} \times R_2^4$$

and where $B = [0, 1]$ and R_1, R_2 define the search domain for the parameters, which usually represents an admissible region, e.g. ensuring that the weighting functions are stable and of minimum phase.

Case Study A: Optimization of Nominal Plant Specifications with Time Delay $\tau_1 = \tau_2 = 0.5$

This proposed algorithm has been used to satisfy the performance design specification for the nominal plant G_n using the configuration of Fig. 6.2. The design criteria are derived from Eqns. 6.12 and 6.13.

$$\gamma_0(G_n, W_1, W_2) \leq \varepsilon_\gamma \tag{6.35}$$
$$\phi_i(G_n, W_1, W_2) \leq \varepsilon_i \quad \text{for } i = 1, 2, \ldots, 16 \tag{6.36}$$

For stability robustness, the value of ε_γ should not be too large, and is here taken as

$$\varepsilon_\gamma = 5.0 \tag{6.37}$$

The performance functionals $\phi_i(G_n, W_1, W_2)$ and the respective prescribed bounds are decided from the design specifications and are shown in second column of Table 6.3. The parameters of the HGA used in the simulation are tabulated in Table 6.1.

Table 6.1. Parameter setting of HGA

Population Size		40
Generation Gap[†]		0.2
Control Gene		
Resolution		1 bit
Range	B	[0, 1]
Crossover	1-point Crossover	Crossover Rate = 0.7
Mutation	Bit Mutation	Mutation Rate = 0.05
Coefficient Gene		
Resolution		10 bits
Range	R_1	(0,2)
	R_2	(0,500)
Crossover	3-point Crossover	Crossover Rate = 0.8
Mutation	Bit mutation	Mutation Rate = 0.1

It took about 135 generations to obtain the optimal compensators. The weighting functions obtained were:

$$W_1 = \frac{(s+1.2800)(s+1.5005)}{s(s+0.8215)(s+1.4868)} \begin{bmatrix} 2.4390 & 0 \\ 0 & 5.8533 \end{bmatrix}$$

$$W_2 = \frac{(s+1.7873)(s^2+0.5620s+1.9844)}{(s+1.7385)(s^2+1.4946s+1.8517)} \begin{bmatrix} 36.0976 & 0 \\ 0 & 36.5854 \end{bmatrix}$$

with $\gamma_0 = 3.6147$ which successfully satisfy Eqns. 6.35 and 6.36. The convergence of the objective value is plotted in Fig. 6.3.

Extreme plants G_1, G_2, G_3, G_4 with system parameters shown in Table 6.2 were used for testing the system's robustness. These extreme plant models were judged to be the most difficult to obtain simultaneous good performance, and it was found that the final system was not very robust.

[†] Number of New Chromosomes Generated = Generation Gap × Population Size

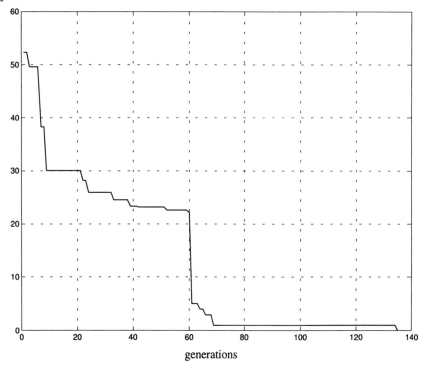

Fig. 6.3. Objective value vs generations

Table 6.2. Extreme plants G_j for $j = 1, 2, 3, 4$

	τ_1	τ_2	k_1	k_2
G_1	0	0	0.8	0.8
G_2	1	1	0.8	1.2
G_3	1	1	1.2	0.8
G_4	1	1	1.2	1.2

Case Study B: Optimization of Plant Specifications with Time Delay $\tau_1 = \tau_2 = 1$

The setting is the same as that for *Case study A*, except that the design criteria, Eqn. 6.36, are modified. This realizes the design criteria where $\tau_1 = \tau_2 = 1$ which is considered more difficult to achieve. The design criteria are modified as follows:

$$\gamma_0(G_n, W_1, W_2) \leq \varepsilon_\gamma$$
$$\phi_i(G_m, W_1, W_2) \leq \varepsilon_i \quad for\ i = 1, 2, \ldots, 16 \qquad (6.38)$$

where G_m is the plant with $k_1 = k_2 = 1$ and $\tau_1 = \tau_2 = 1$ using a fifth-order Padé approximation.

It took about 800 generations to obtain W_1 and W_2 for robust feedback control. The parameters of W_1 and W_2 were

$$W_1 = \frac{(s+0.8878)(s^2+0.3161s+1.1044)}{s(s+1.1278)(s^2+1.1356s+0.1444)} \begin{bmatrix} 13.1707 & 0 \\ 0 & 13.1707 \end{bmatrix}$$

$$W_2 = \frac{(s+0.1678)(s^2+0.3727s+0.7161)}{(s+1.5727)(s^2+1.0049s+1.8946)} \begin{bmatrix} 50.2439 & 0 \\ 0 & 52.1951 \end{bmatrix}$$

where $\gamma_0 = 3.3047$.

The closed-loop performances are tabulated in Table 6.3 and depicted in Fig. 6.4. All the design criteria are satisfied except that the 50dB gain limit is marginally exceeded by $\phi_{15}(G_2)$ and $\phi_{15}(G_3)$.

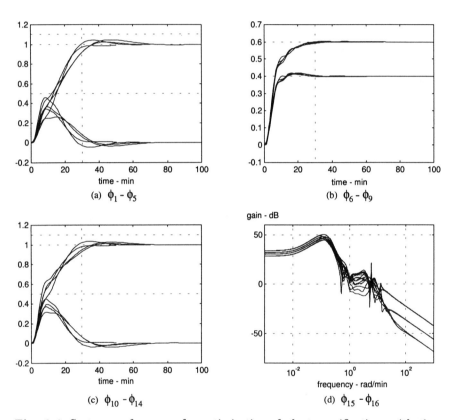

Fig. 6.4. System performance for optimization of plant specifications with time delay $\tau_1 = \tau_2 = 1$

6.1 A Mixed Optimization Design Approach   145

Table 6.3. Final system performance for optimization of plant specifications with time delay $\tau_1 = \tau_2 = 1$

i	ε_i	$\phi_i(G_m)$	$\phi_i(G_1)$	$\phi_i(G_2)$	$\phi_i(G_3)$	$\phi_i(G_4)$
1	1.1	1.0212	1.0410	1.0377	1.0306	1.0030
2	-0.9	-0.9732	-0.9655	-0.9898	-0.9870	-0.9687
3	0.5	0.3556	0.3625	0.2844	0.4290	0.3471
4	0.1	0	0	0.0001	0.0002	0
5	0.1	0.0159	0.0305	0.0232	0.0477	0.0054
6	0.5	0.4177	0.4198	0.4272	0.4274	0.4308
7	-0.35	-0.3981	-0.3966	-0.3851	-0.3873	-0.3843
8	0.7	0.6012	0.6023	0.6149	0.6162	0.6191
9	-0.55	-0.5972	-0.5967	-0.5838	-0.5873	-0.5777
10	0.5	0.3837	0.3912	0.4300	0.3359	0.3746
11	1.1	1.0124	1.0242	1.0299	1.0372	1.0138
12	-0.9	-0.9832	-0.9786	-0.9748	-0.9841	-0.9660
13	0.1	0.0171	0.0330	0.0417	0.0270	0.0058
14	0.1	0	0	0.0003	0.0002	0
15	50.0	48.4479	49.7050	51.0427	51.0167	47.76
16	150.0	9.0773	10.7159	9.8627	10.000	9.8627

Case Study C: Optimization of Overall Plants Specifications with Extreme Conditions

Since it may not be easy to obtain a controller that satisfies the performance specifications for extreme plant models by optimization of the nominal plant or a typical plant, an alternative will simultaneously optimize all of the extreme plants. The design criteria are now re-defined as

$$\gamma_0(W_1, W_2) \leq \varepsilon_\gamma \quad (6.39)$$
$$\phi_i(G_j, W_1, W_2) \leq \varepsilon_i \quad \text{for } i = 1, 2, \ldots, 16; \; j = 1, 2, 3, 4 \quad (6.40)$$

A Multiple Objective HGA (MO-HGA) has been applied here. A multiple objective ranking [63] approach is used. The chromosome I is ranked as

$$rank(I) = 1 + p \quad (6.41)$$

if I is dominated by other p chromosomes in the population.

From Eqn. 6.40, 64 objectives need to be achieved. Such huge numbers of objectives demand a large number of comparison operations. Hence, these have been simplified into 4 objectives to indicate the fitness for each extreme plant as before. Define m_{ij} for extreme plant $i = 1, 2, \ldots, 16$ and $j = 1, 2, 3, 4$ as

$$m_{ij} = \begin{cases} 0 & \text{if } \phi_i(G_j, W_1, W_2) \leq \varepsilon_i \\ 1 & \text{else} \end{cases} \quad (6.42)$$

The objective f_j for extreme plant j, for $j = 1, 2, 3, 4$, is

6. Genetic Algorithms in H-infinity Control

$$f_j = \begin{cases} \sum_{i=1}^{n} m_{ij} & \text{if } \gamma_0 < \varepsilon_\gamma \\ n+1+\gamma_0 & \text{else} \end{cases} \quad (6.43)$$

where $n = 16$.

After 448 generations, W_1 and W_2 were obtained and expressed as follows:

$$W_1 = \frac{(s+0.8956)(s^2+0.7161s+1.4888)}{s(s+1.7249)(s^2+1.9122s+0.1444)} \begin{bmatrix} 76.0976 & 0 \\ 0 & 47.3171 \end{bmatrix}$$

$$W_2 = \frac{(s+1.4537)(s^2+0.1990s+0.5444)}{(s+1.4498)(s^2+1.2741s+1.9551)} \begin{bmatrix} 17.0732 & 0 \\ 0 & 17.5610 \end{bmatrix}$$

with $\gamma_0 = 3.1778$. Fig. 6.5 demonstrates the multiple objective optimization process of the proposed MO-HGA. Trade-offs between different objective values can be noticed.

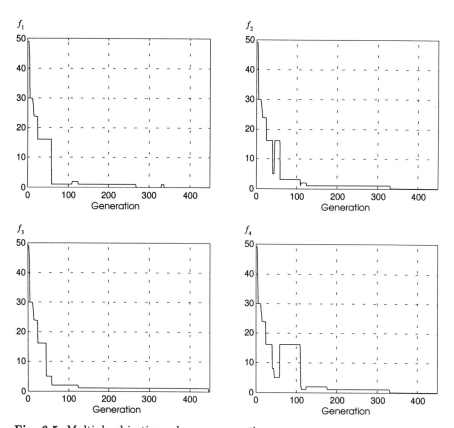

Fig. 6.5. Multiple objective values vs generations

The closed loop system responses for the extreme plants are tabulated in Table 6.4 and depicted in Fig. 6.6.

Table 6.4. Final system performance for optimization of overall plants specifications with extreme conditions

i	ε_i	$\phi_i(G_1)$	$\phi_i(G_2)$	$\phi_i(G_3)$	$\phi_i(G_4)$
1	1.1	1.0477	1.0142	1.0401	1.0024
2	-0.9	-0.9159	-0.9791	-0.9125	-0.9525
3	0.5	0.3843	0.3204	0.4193	0.3578
4	0.1	0	0.0001	0.0001	0.0001
5	0.1	0.0363	0.0168	0.0392	0.0018
6	0.5	0.4266	0.4242	0.4251	0.4218
7	-0.35	-0.3960	-0.3967	-0.3963	-0.3998
8	0.7	0.6030	0.6023	0.6034	0.6003
9	-0.55	-0.5938	-0.5915	-0.5976	-0.5969
10	0.5	0.4065	0.4604	0.3569	0.3786
11	1.1	1.0292	1.0128	1.0315	1.0015
12	-0.9	-0.9478	-0.9409	-0.9921	-0.9709
13	0.1	0.0384	0.0139	0.0328	0.0019
14	0.1	0	0.0002	0.0001	0.0002
15	50.0	48.8978	49.5575	49.7752	45.8310
16	150.0	16.9133	13.1862	11.6430	11.9696

6.1.3 Application II: Benchmark Problem

The investigated benchmark process [85] aims to find the controllers for the time-varying behaviour of the plant which exhibits three different stress levels (S_1, S_2 and S_3). S_3 is signified as the highest stress level that induces the largest variations, whereas S_1 is the lowest and S_2 is in the medium range. The overall plant model that comprises the dominant dynamics and the high-frequency dynamics is shown as below:

$$G_{bp}(s) = G(s) \cdot G_{hf}(s) \tag{6.44}$$

where the dominant dynamics, i.e. the nominal plant

$$G(s) = \frac{K(-T_2 s + 1)\omega_0^2}{(s^2 + 2\zeta\omega_0 s + \omega_0^2)(T_1 s + 1)} \tag{6.45}$$

and the high frequency dynamics

$$G_{hf}(s) = \frac{\omega_\delta^2}{(s^2 + 2\zeta_\delta\omega_\delta s + \omega_\delta^2)(T_1^\delta s + 1)(T_2^\delta s + 1)} \tag{6.46}$$

The high-frequency dynamics are fixed in time, taking the values of $T_1^\delta = 1/8$, $T_2^\delta = 1/12$, $\omega_\delta = 15$, and $\zeta_\delta = 0.6$ for all stress levels.

148   6. Genetic Algorithms in H-infinity Control

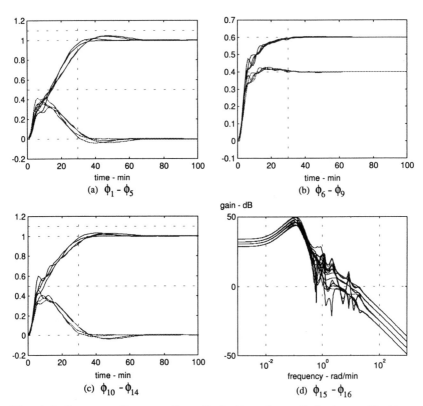

Fig. 6.6. System performance for optimization of overall plants specifications with extreme conditions

6.1 A Mixed Optimization Design Approach

The dominant dynamics is time-varying within the limited ranges: $T_1 = 5 \pm \Delta T_1$, $T_2 = 0.4 \pm \Delta T_2$, $\omega_0 = 5 \pm \Delta \omega_0$, $\zeta = 0.3 \pm \Delta \zeta$ and $K = 1 \pm \Delta K$, where the ranges of variation for different stress levels are tabulated in Table 6.5. A

Table 6.5. Plant parameters variation at three stress levels

	S_1	S_2	S_3
ΔT_1	0.2	0.3	0.3
ΔT_2	0.05	0.1	0.15
$\Delta \omega_0$	1.5	2.5	3
$\Delta \zeta$	0.1	0.15	0.15
ΔK	0	0.15	0.5

square wave oscillating signal with a strength between ± 1.0 for a period of 20 seconds is considered as the reference signal for excitation. For all three stress levels, the plant should respond to the reference as quickly as possible, but subject to the following closed-loop performance constraints:

1. Plant output values *must* be between ± 1.5 *at all times*;
2. The overshoot and undershoot should preferably be less than 0.2 *most of the time*, though occasionally larger overshoot or undershoot are acceptable provided that the ± 1.5 limits are preserved;
3. Control input to the plant saturates at ± 5;
4. Fast settling time; and
5. Zero steady state tracking error (modulo high frequency noise).

According to the performance criteria of the benchmark, four performance indices, namely overshoot (ϕ_1), undershoot (ϕ_2), settling time (ϕ_3) and rising time (ϕ_4), are defined as:

$$\begin{aligned} \phi_1 &= \max(y) \\ \phi_2 &= -\min(y) \\ \phi_3 &= \max_t(|y - r| > 0.05) \\ \phi_4 &= \min_t(|y| \geq |0.8r|)^\ddagger \end{aligned} \quad (6.47)$$

where r and y are the reference signal and plant output, respectively.

Various methods such as PI, pole placement, adaptive control design in the form of model reference adaptive control and adaptive predictive control [85], have been carried out to address the benchmark problem. A LSDP maximizing a $\mathbf{H}^\infty$-norm, combined with the method of inequalities (MOI), was also adopted to address this problem [85, 235]. All of these appear have

---

$\ddagger$ Based on the definition in [161] but with some modifications for simplicity

reached some degree of success to the extent that the prescribed performance can be achieved.

To demonstrate the effectiveness of this approach, the system variations tabulated in Table 6.5 were examined. Eight extreme plant models (G_i, i=1, 2, ..., 8) were then selected and listed in Table 6.6.

Table 6.6. Extreme plant models

	T_1	T_2	ω_0	ζ	K
G_1	$5 + \Delta T_1$	$0.4 + \Delta T_2$	$5 + \Delta\omega_0$	$0.3 + \Delta\zeta$	$1 + \Delta K$
G_2	$5 + \Delta T_1$	$0.4 + \Delta T_2$	$5 + \Delta\omega_0$	$0.3 - \Delta\zeta$	$1 - \Delta K$
G_3	$5 + \Delta T_1$	$0.4 - \Delta T_2$	$5 + \Delta\omega_0$	$0.3 - \Delta\zeta$	$1 + \Delta K$
G_4	$5 + \Delta T_1$	$0.4 - \Delta T_2$	$5 - \Delta\omega_0$	$0.3 - \Delta\zeta$	$1 + \Delta K$
G_5	$5 - \Delta T_1$	$0.4 + \Delta T_2$	$5 + \Delta\omega_0$	$0.3 - \Delta\zeta$	$1 - \Delta K$
G_6	$5 - \Delta T_1$	$0.4 - \Delta T_2$	$5 - \Delta\omega_0$	$0.3 + \Delta\zeta$	$1 + \Delta K$
G_7	$5 + \Delta T_1$	$0.4 - \Delta T_2$	$5 - \Delta\omega_0$	$0.3 - \Delta\zeta$	$1 - \Delta K$
G_8	$5 - \Delta T_1$	$0.4 - \Delta T_2$	$5 - \Delta\omega_0$	$0.3 - \Delta\zeta$	$1 - \Delta K$

A population of size 40 is randomly generated initially. At each MO-HGA cycle, eight offspring are generated through crossover and mutation. The parameters of the crossover and mutation operations are tabulated in Table 6.7. After 1,000 iterations, the final weighting functions W_1 and W_2 are obtained as indicated in Table 6.8.

Table 6.7. Parameters of crossover and mutation

	Control Gene	Coefficient Gene
Crossover	1-point crossover	3-point crossover
Crossover rate	0.7	0.8
Mutation	bit mutation	bit mutation
Mutation rate	0.05	0.1

Table 6.8. Resulted weighting functions for three stress levels

	W_1	W_2
S_1	$\dfrac{7.32(s+0.09)}{s(s+1.50)}$	1
S_2	$\dfrac{5.85(s+0.15)}{s(s+1.69)}$	1
S_3	$\dfrac{8.29(s+0.80)(s^2+0.76s+1.97)}{s(s+1.75)(s^2+1.97s+1.77)}$	$\dfrac{s+0.69}{s+1.57}$

6.1 A Mixed Optimization Design Approach

Simulation work proceeded to justify the closed loop system performance. This was done by performing the process repeatedly 15 times in each of the three stress levels, so that a much clearer picture of the plant dynamics variation can be represented. The simulated results obtained for each stress level are shown in Figs. 6.7–6.9. It can be seen that these results indicate that the design specifications have been achieved despite the system variation of the stress levels.

By comparison, the MO-HGA design approach based on $\mathbf{H}^\infty$ LSDP formulation provides an even better simplicity versus performance benefit for small to moderate plant variations [85]. Furthermore, in contrast to the robust $\mathbf{H}^\infty$ design [229], the present method of controlling the plant dynamics in stress level 3 is also much improved and the obtained controller is now simplified. It should be noted that the pre-defined order and structure of the weighting functions are not required in this approach. The result indicates that the specification meets all the design criteria and may be considered a compatible alternative to the properly tuned adaptive controller in stress level 3 [85].

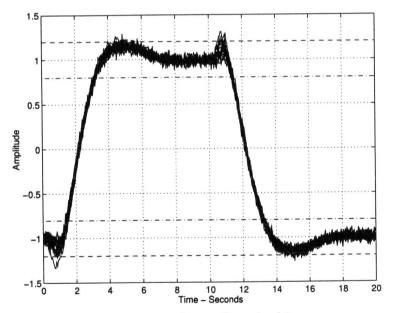

Fig. 6.7. Plant response to square reference: Stress level 1

152   6. Genetic Algorithms in H-infinity Control

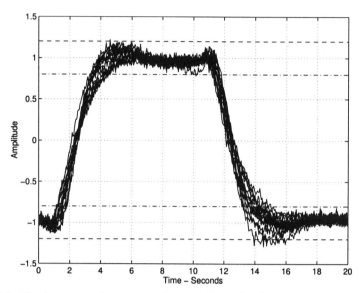

Fig. 6.8. Plant response to square reference: Stress level 2

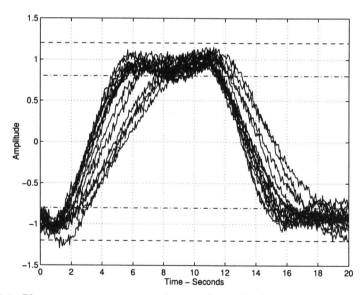

Fig. 6.9. Plant response to square reference: Stress level 3

6.1.4 Design Comments

The proposed HGA enables simultaneous searching of the structures and coefficients of the weighting functions. this is a unique approach for such a design. Several advantages have been gained from this method:

- The HGA can easily handle the constraints to ensure the stability of the weight functions;
- a multiple objective approach can be adopted to address the conflicting control design specifications; and
- the structures of the weighting functions are no longer pre-fixed; but only a fundamental structure is required, which provides the optimality for the solution over several different types of weights.

In the case studies, the performance was evaluated for a selection of extreme plant models chosen by the designer. The problem of efficiently determining the worst-case performance over the range of plants still remains. Since the proposed algorithm follows the formulation of an MOI which requires the choice of several plants only, it is necessary to choose the most representative of all possible plant models.

7. Hierarchical Genetic Algorithms in Computational Intelligence

It is anticipated that future engineering design will relegate its own disciplinary concepts and will become heavily involved with computational intelligence (CI). This trend of development is understandable since computing power has become so much faster and cheaper nowadays, such that a required solution can be automatically obtained even when this is based upon a computationally intensive scheme.

The use of CI is widespread and forms a core unit for the production of this emerging technology that is being utilized in both academic and industrial domains. The noticeable development of neural networks and fuzzy logic systems for engineering applications are just two typical examples that illustrate this point of view. Very often, when CI is applied, and despite the capabilities of the technology, prior knowledge of the system's topology and governing parameters of concern is required in order to fully exploit the use of CI. Unfortunately, this is not easily come by and can sometimes only be achieved by sheer computing power.

Having discovered the attributes of the HGA, in Chap. 5.1, as a means of solving the topological structure of filtering problems, the HGA is also used to solve problems of a similar nature. In this case, the HGA is used to tackle the well known neural network (NN) topology as well as the fuzzy logic membership functions and rule problems. It is believed from the investigations that the HGA offers a sound approach to reaching an optimal but reduced topology in both cases It is our belief that the HGA will become a potent technology, and that this trend should be encouraged for future system design.

7.1 Neural Networks

The use of NN for system control and signal processing has been well accepted. The most noticeable applications are in the areas of telecommunication, active noise control, pattern recognition, prediction and financial analysis, process control, speech recognition, etc. [5, 237]. The widespread use of NN is due mainly to its behavioural emulation of the nature of the human brain and the fact that its structure can be mathematically formulated. An efficient NN topology is capable of enhancing the system performance in terms of learning speed, accuracy, noise resistance and generalization ability. Hence, NN can be considered as a parallel and distributed processing unit that consists of multiple processing elements. This structure is complex, unknown and must be pre-defined initially.

To this end, the required technique to obtain an optimal NN topology can be generalized into:

1. a prior analysis of the potentialities of a network;
2. optimum size of a network that reduces the enormous search spaces in learning, (to improve the computational power);
3. use of a computationally prohibitive construction-destruction algorithm [30, 162]; and
4. use of mathematical methods to determine the architecture and parameters of the network [244].

The recent development of CI to explore a different approach to optimize the network configuration has caused great interest in the NN area. The application of fuzzy logic techniques to adapt the network configuration [196] has been reported. Evolutionary Algorithms have been applied to determine the network construction and/or the connection weight optimization [4, 150, 152]. The promising results obtained by using GA [143, 167] to train and design artificial neural networks has proved to be a useful technique for integration. All these techniques provide viable alternatives for the improvement of NN topology.

In this section, the HGA is proposed for the optimization of NN topology. The advantage of this approach is that the genes of the chromosome are classified into two categories in a hierarchical form. This is an ideal formulation for the genes as the layers, neurons, connection weightings and bias can be formed within the string of chromosomes. This provides a vast dimension for genetic operation which, in the end, improves the computational power and the optimization of NN topology.

7.1.1 Introduction of Neural Network

A general neural network topology takes the form of a multilayer feedforward network as depicted in Fig. 7.1.

The basic processing element is called a neuron. A neuron consists of an activity level, a set of input and output connections with a bias value associated to each connection. Fig. 7.2 shows a typical neuron with n-input

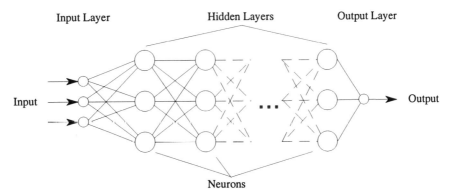

Fig. 7.1. A multilayer feedforward neural network topology

connections and a single output connection.

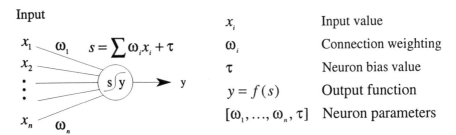

x_i	Input value
ω_i	Connection weighting
τ	Neuron bias value
$y = f(s)$	Output function
$[\omega_1, ..., \omega_n, \tau]$	Neuron parameters

Fig. 7.2. A single neuron

The output of the neuron is determined as:

$$y = f\left(\sum_{i=1}^{n} \omega_i x_i + \tau\right) \quad (7.1)$$

where $x_1, x_2, \ldots, x_n$ are input signals; $\omega_1, \omega_2, \ldots, \omega_n$ are connection weightings; τ is the bias value; and f is a defined output function that may be a sigmoid, tanh, step function etc.

In order for this topology to function according to design criteria, a learning algorithm that is capable of modifying the network parameters, as indicated in Eqn. 7.1, is of paramount important to the NN. The backpropagation (BP) technique employs a gradient descent learning algorithm [200] that is commonly used by the NN community. This approach suffers from a pre-defined topology such that the numbers of neurons and connections must be known a prior. Furthermore, as the network complexity increases, the performance of BP decreases rapidly. The other deficit of BP is its use of gradient search algorithms, where discontinuous connection weightings cannot be handled.

7.1.2 HGA Trained Neural Network (HGANN)

Having realized the pros and cons of NN, the bottle-neck problem lies within the optimization procedures that are implemented to obtain an optimal NN topology. Hence, the formulation of the HGA is applied for this purpose [218]. The HGA differs from the standard GA with a hierarchy structure in that each chromosome consists of multilevels of genes.

Fig. 7.3 shows the chromosome representation in the HGANN system. Each chromosome consists of two types of genes, i.e. control genes and connection genes. The control genes in the form of bits, are the genes for layers and neurons for activation. The connection genes, a real- value representation, are the genes for connection weightings and neuron bias. A neural network defined by this structure is depicted in Fig. 7.3.

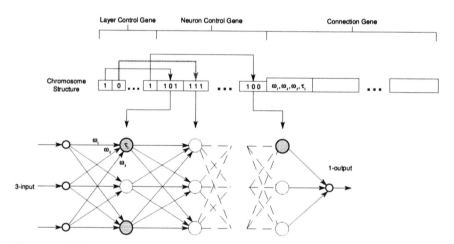

Fig. 7.3. HGANN chromosome structure

Within such a specific treatment, a structural chromosome incorporates both active and inactive genes. It should be noted that the inactive genes remain in the chromosome structure and can be carried forward for further generations. Such an inherent genetic variation in the chromosome avoids any trapping at local optima which has the potential to cause premature convergence. Thus it maintains a balance between exploiting its accumulated knowledge and exploring the new areas of the search space. This structure also allows larger genetic variations in chromosome while maintaining high viability by permitting multiple simultaneous genetic changes. As a result, a single change in high level genes will cause multiple changes (activation or deactivation in the whole level) in lower level genes. In the case of the traditional GA, this is only possible when a sequence of many random changes takes place. Hence the computational power is greatly improved.

To formulate such an HGANN, its overall system block diagram is shown in Fig. 7.4. This hierarchical genetic trained neural network structure has the ability to learn the network topology and the associated weighting connection concurrently. Each learning cycle is known as a generation.

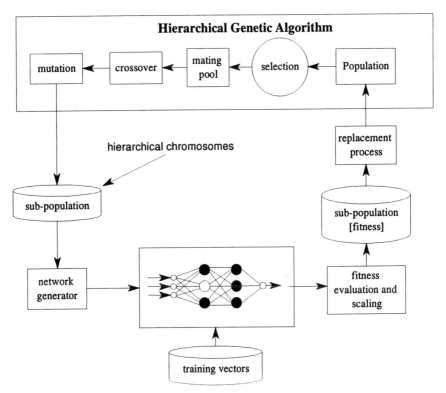

Fig. 7.4. Block diagram of the overall HGANN operation

Population. In order to explore all possible topologies, the population of HGANN at k-th generation, $P^{(k)}$, is divided into several connection subgroups,

$$G_1^{(k)} \cup G_2^{(k)} \ldots \cup G_M^{(k)} = P^{(k)} \quad \text{and}$$
$$G_i^{(k)} \cap G_j^{(k)} = \phi, \quad \forall i \neq j \tag{7.2}$$

where M is the maximum number of possible connections represented by HGANN; and $G_i^{(k)}$ is the subgroup of chromosomes that represents those networks with i active connection at k-th generation.

A concurrent factor, λ, is used to define the maximum number of chromosome stored in each sub-group,

$$size\left[G_i^{(k)}\right] \leq \lambda \tag{7.3}$$

where $size\left[G_i^{(k)}\right]$ is the number of elements in $G_i^{(k)}$.

Hence, the maximum population size is limited to P_{max} which is defined as

$$P_{max} = \lambda M \tag{7.4}$$

In the initialization stage, there are $P^{(0)} \leq P_{max}$ chromosomes to be generated. Once a new generation has been produced, new chromosomes are inserted into the population pool to explore the next possible generation of topology.

Objective Functions. The objective of training the network is to minimize two different parameters: the accuracy of the network (f_1) and the complexity of the network (f_2) which is simply defined by the number of active connections in the network. This is calculated based upon the summation of the total number of active connections taking place. The accuracy of the network (f_1) is defined as:

$$f_1 = \frac{1}{N}\sum_{i=1}^{N}(\hat{y}_i - y_i)^2 \tag{7.5}$$

wher N is the size of the testing vector; $\hat{y}_i$ and y_i are the network output and desired output for the i-th pattern of the test vector respectively.

Selection process. Parent Selection is a routine to emulate the survival-of-the-fittest mechanism of nature. Chromosomes in the population are selected for the generation of new chromosomes (offspring) by the certain selection schemes. It is expected that a better chromosome will receive a higher number of offspring and thus has a higher chance of surviving in the subsequent generation. Since there are two different objective functions, (f_1) and (f_2) of the network optimization process, the fitness value of chromosome z is thus determined:

$$f(z) = \alpha \cdot rank[f_1(z)] + \beta \cdot f_2(z) \tag{7.6}$$

where α is accuracy weighting coefficient; β is complexity weighting coefficient; and $rank[f_1(z)] \in Z^+$ is the rank value.

The selection rate of a chromosome z, $tsr(z)$, is determined by:

$$tsr(z) = \frac{F - f(z)}{(size[P^{(k)}] - 1) \cdot F} \quad (7.7)$$

where F is the sum of the fitness value of all chromosomes.

Considering that the accuracy of the network is of paramount importance rather than the complexity of the network, the rule of thumb of the design is such that the weighting coefficients, α, and β take the form as follows:

Let M be the maximum active number of connections in the neural network system, then

$$f_2(z) \leq M, \quad \forall z \in P \quad (7.8)$$

Assuming that at least one successful network has been learnt in the population P, i.e. $\exists z_i \in P$, such that $f_1(z_i) = 0$ and $rank[f_1(z_i)] = 1$, then

$$\begin{aligned} f(z_i) &= \alpha + \beta \cdot f_2(z_i) \\ &\leq \alpha + \beta \cdot M \end{aligned} \quad (7.9)$$

where $\beta \in \Re^+$.

Consider that chromosome $z_j \in P$ is failed in learning, i.e. $f_1(z_j) > 0 \Rightarrow rank[f_1(z_j)] \geq 2$,

$$\begin{aligned} f(z_j) &= \alpha \cdot rank[f_1(z_j)] + \beta \cdot f_2(z_j) \\ &\geq 2\alpha + \beta \cdot f_2(z_j) \\ &> 2\alpha \end{aligned} \quad (7.10)$$

Hence, α is set as following to ensure $f(z_j) > f(z_i)$,

$$\alpha > \beta \cdot M \quad (7.11)$$

Genetic Operations. Since there are two types of genes in the chromosome structure, see Fig. 7.3, specific genetic operations have been designed to suit their purposes. For each type of gene, there are two genetic operations, i.e. crossover and mutation which are recommended.

Control Genes Crossover. A modified multilayer one point crossover operation is applied into the control genes with the probability rate p_{cb}. Once the probability test has passed (a randomly generated number, r_1, is smaller than p_{cb}), one-point crossover is performed in each layer as shown in Fig. 7.5. Parents are separated into two portions by a randomly defined crosspoint at each level. The new control genes are then formed by combining the first part of the parent 1 and the second part of the parent 2 as indicated in Fig. 7.5.

162   7. Hierarchical Genetic Algorithms in Computational Intelligence

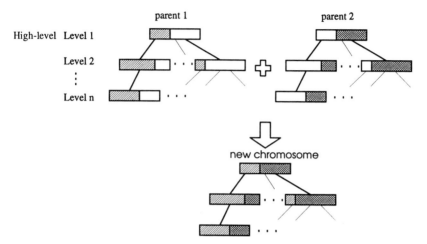

Fig. 7.5. Control genes' multilevel one-point crossover operation

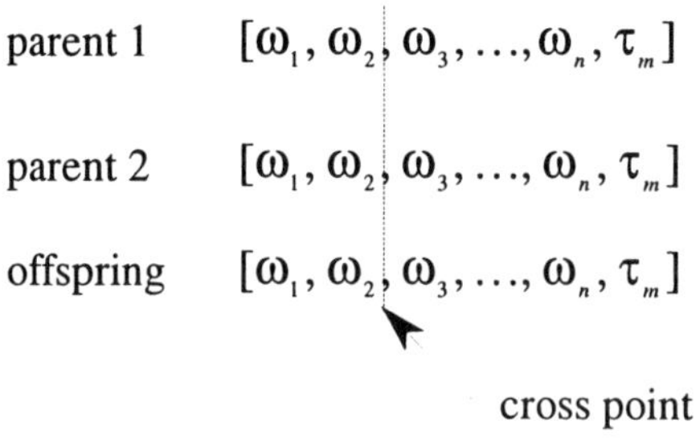

parent 1 $[\omega_1, \omega_2, \omega_3, \ldots, \omega_n, \tau_m]$

parent 2 $[\omega_1, \omega_2, \omega_3, \ldots, \omega_n, \tau_m]$

offspring $[\omega_1, \omega_2, \omega_3, \ldots, \omega_n, \tau_m]$

cross point

Fig. 7.6. Connection genes' crossover operation

Connection Genes Crossover. Since the connection gene is a vector of real parameters, a one-point crossover operation can thus be directly applied. The operation rate is assumed to be p_{cr}. If a randomly generated number, r_2, is smaller than p_{cr}, the new gene is mated from the first portion of the parent 1 and the last portion in the parent 2 as shown in Fig. 7.6.

Control Genes' Mutation. Bit Mutation is applied for the control genes in the form of a bit-string. This is a random operation that occasionally (with probability p_{mb}, typically 0.01-0.05) occurs which alters the value of a string position so as to introduce variations into the chromosome. Each bit of the control gene is flipped if a probability test is satisfied (a randomly generated number, r_3, is smaller than p_{mb}). An example of Control Genes' Mutation is demonstrated in Fig. 7.7.

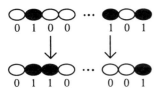

Fig. 7.7. Connection genes' mutation operation

Connection Genes mutation. A real value mutation has been designed for the connection genes. For each connection gene, a Gaussian noise is added with probability p_{mr} which can be randomly assigned, (typically 0.05-0.1). The new mutation function is thus:

$$m_r(x) = x + N(0,1) \qquad (7.12)$$

where x is the current connection weight, and $N(\mu, \sigma^2)$ is a Gaussian random variable with mean μ and variance σ^2.

Insertion Strategy. The top level description of the insertion strategy for the new chromosome z is expressed in Table 7.1.

7.1.3 Simulation Results

To verify the performance of the proposed HGANN system, a subset of suggesting testing functions (a and b) in [184] and an arbitrarily logic function (c) to assess the NN have been used. The following three different 3-input Boolean functions have been introduced for the verification of the HGANN:

(a) Test 1 : XOR
(b) Test 2 : Parity check
(c) Test 3 : Arbitrarily set Logic Function

Table 7.1. Insertion strategy

Step 1:	If $\{G^{(k)}_{f_2(z)} = \phi$ or $size\left[G^{(k)}_{f_2(z)}\right] < \lambda\}$
	then
	$\quad \{G^{(k+1)}_{f_2(z)} = G^{(k)}_{f_2(z)} \cup \{z\}$ and goto step 3 $\}$
	else
	$\quad$ goto step 2
Step 2:	If $\{f_1(z) < f_m = max\{f_1(z_i), \forall z_i \in G^{(k)}_{f_2(z)}\}\}$
	then
	$\quad \{G^{(k+1)}_{f_2(z)} = \{z_i : f_1(z_i) < f_m, z_i \in G^{(k)}_{f_2(z)}\} \cup \{z\}\}$
	else
	$\quad$ goto step 3
Step 3:	Exit

Table 7.2. Parameters for genetic operations of HGANN

Population Size	20
Generation Gap	1.0
Selection	Roulette Wheel Selection on Rank
Reinsertion	Table 7.1

Table 7.3. Parameters for chromosome operations of HGANN

	Control Genes	Connection Genes
Representation	Bit Representation (1 bit)	Real Number
Crossover	One point Crossover	One point Crossover
Crossover Rate	1.0	1.0
Mutation	Bit Mutation	Random Mutation
Mutation Rate	0.05	0.1

$$input = \{x_1, x_2, x_3\}; \quad output = x_2 \wedge (x_1 \vee x_3)$$

The genetic operational parameters are shown in Tables 7.2 and 7.3.

For each of above three test cases, 30 trials were conducted to allow comparison of the performance for a single layer GA* (GA) and Back-Propagation (BP) [67] against the HGANN. The basic topology for learning is depicted in Fig. 7.8. It should be noticed that BP was applied to this base topology for all three tests. In the case of the GA and HGANN, both the topology and the connection weights were optimized. The number of chromosome levels and the gene length for HGANN were both set to two. Simulation results are tabulated in Tables 7.4-7.6.

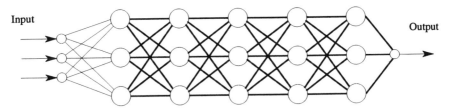

Fig. 7.8. Basic topology for learning

Table 7.4. Mean of terminated iterations in 30 trials

Test	HGANN	GA	BP
1	513	1121	1652 [1][†]
2	870	2228	1354, [2]
3	37	57	293, [0]

Table 7.5. Best of terminated iterations in 30 trials

Test	HGANN	GA	BP
1	38	134	359
2	187	435	279
3	5	6	220

The medium simulation results of the three tests for different algorithms are depicted in Figs. 7.9-7.11. The corresponding topologies obtained after

---

* Single layer GA has a chromosome structure without any layer control genes.
[†] Number of trial failed in network training after 10,000 iteration. Terminated iteration number 10,000 is assigned.

166   7. Hierarchical Genetic Algorithms in Computational Intelligence

Table 7.6. Standard deviation of terminated iterations in 30 trials

Test	HGANN	GA	BP
1	431	652	1782
2	870	869	2345
3	28	52	84

1000 iterations are shown in Figs. 7.12-7.14. It can be observed that the HGANN has a faster convergence rate than the other two methods for all three simulation tests. Moreover, the number of connections was also minimized concurrently in HGANN.

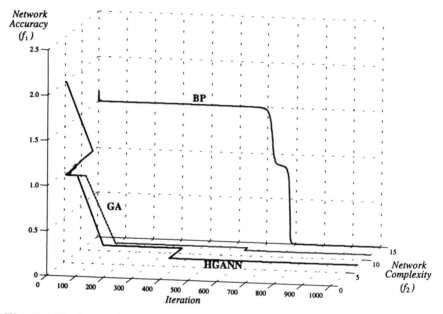

Fig. 7.9. Median performance in test 1

This new scheme has the ability to optimize the NN topology concurrently so that it includes the number of layers and neurons, as well as the associated connection weightings. As a result, the final obtained NN is optimal and is considered to be much more effective for tackling practical problems. Judging from the simulation results, this new computational architecture gives a better performance than techniques using BP and traditional GA methods in terms of the number of training iterations.

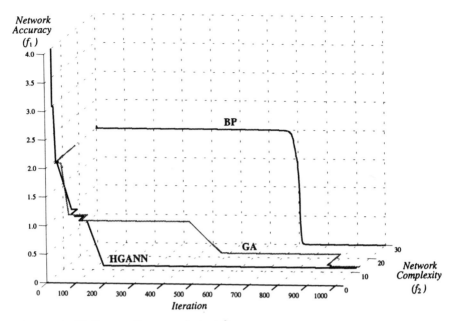

Fig. 7.10. Median performance in test 2

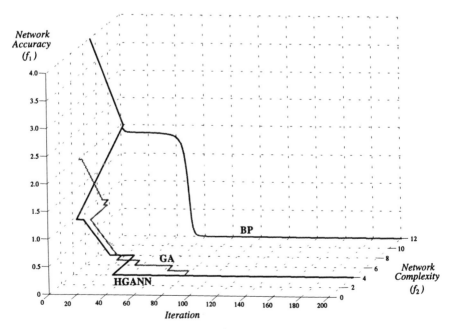

Fig. 7.11. Median performance in test 3

168   7. Hierarchical Genetic Algorithms in Computational Intelligence

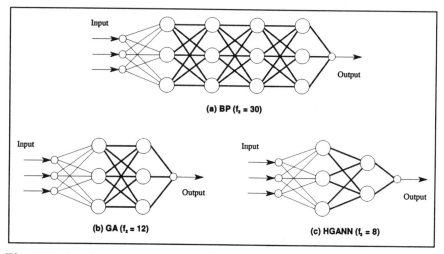

Fig. 7.12. Topology for test 1 after 1000 iterations

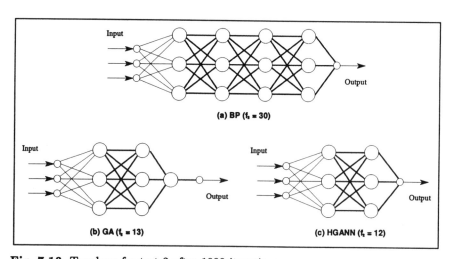

Fig. 7.13. Topology for test 2 after 1000 iterations

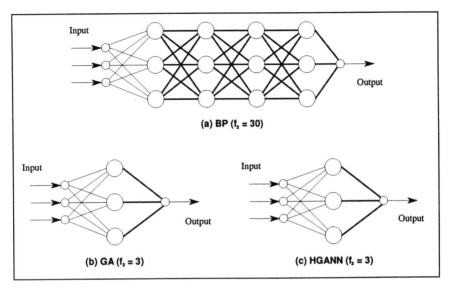

Fig. 7.14. Topology for test 3 after 1000 iterations

7.1.4 Application of HGANN on Classification

Despite the successful demonstrations as shown above, using the benchmark problems that were often used in the 1980s, these results are not convincing enough to justify this approach for practical implementation. A number of deficiencies that have to be overcome for a realistic application exist:

1. all of these problems are purely synthetic and have strong prior regularities in their structure;
2. For some, it is unclear how to measure these in a meaningful way according to the generalization capabilities of a network with respect to the problem.
3. Most of the problems can be solved absolutely, which is untypical for realistic settings.

Therefore, in order to fully justify the capability of the HGANN, a real-life application is introduced. The HGANN technique is used to determine decisions about breast cancer, so that a correct diagnosis can be made for the classification of a tumor as either being benign or malignant.

To facilitate this exercise, the original data was obtained from the University of Wisconsin Hospitals, Madison [136]. There are nine input attributes and two outputs. The input attributes are based on cell descriptions gathered by microscopic examination, of which are listed in Table 7.7.

The data in continuous form were rescaled, first by a linear function so that a mapping into the range of 0...1 could be made. There are 699 examples in total where the class distribution is as follows:

Table 7.7. Input attributes and their domain

Attribute	Domain
Clump Thickness	0-10
Uniformity of Cell Size	0-10
Uniformity of Cell Shape	0-10
Marginal Adhesion	0-10
Single Epithelial Cell Size	0-10
Bare Nuclei	0-10
Bland Chromatin	0-10
Normal Nucleoli	0-10
Mitoses	0-10

Table 7.8. Class distribution

Class	benign	malign	total
Total number	458	241	699
Total percentage	65.5	34.5	100

In real life, the collected data may comprise some missing attribute values. But they are filled by fixed values which were taken to be the mean of the non-missing values of this attribute. Within this data set, there are 16 missing values for 6-th attribute. So here, these are simply encoded as 0.3, since the average value of the attributes is roughly 3.5.

To use these data for the learning of the neural network learning algorithms, the data must be split into at least two parts:

1. one part for NN training, called the *training data*; and
2. another part for NN performance measurement, called the *test set*.

The idea is that the performance of a network on the test set estimates its performance for real use. It turns out that there is absolutely no information about the test set examples or the test set performance of the network available during the training process; otherwise the benchmark is invalid. The data are, now, classified as

1. 100 training data
2. 599 testing data

The fitness function is defined as:

$$f = \sum_{k=1}^{100} |y_t(k) - r_t(k)| \qquad (7.13)$$

where $y_t(k)$ and $r_t(k)$ are the network output and expected output of the training data, respectively.

The error measure (g) is defined as:

$$g = \sum_{k=1}^{599} |y_s(k) - r_s(k)| \qquad (7.14)$$

where $y_s(k)$ and $r_s(k)$ are the network output and expected output of the testing data, respectively.

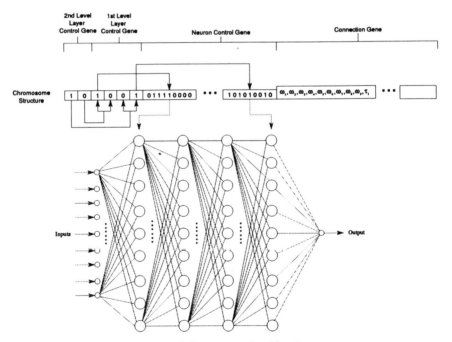

Fig. 7.15. Fundamental network for cancer classification

To proceed with the HGANN formulation, we can design the fundamental network as shown in Fig. 7.15. This is a NN architecture that has nine-input and one-output. The "1" signifies the output as being benign. By the use of the similar GA operational parameters as before, the performance of

HGANN due to the number of identified active neurons is obtained as shown in Fig. 7.16. It can be clearly seen that only four neurons would be necessary instead of 36 neurons (a full connected network) as before. This is a greatly simplified network which is depicted in Fig. 7.17. Based on the error measurement where $g = 26.9$ was obtained, the final HGANN topology will have the accuracy of about 95.5%. This is a slightly better comparable result to that which has already been obtained (93.5% and 93.7%) by [239, 247]. However, in this case, this simple architecture will be much more computationally efficient as compared with the original network (Fig. 7.15). As a result, this could potentially lead to a faster procedure for breast cancer diagnosis.

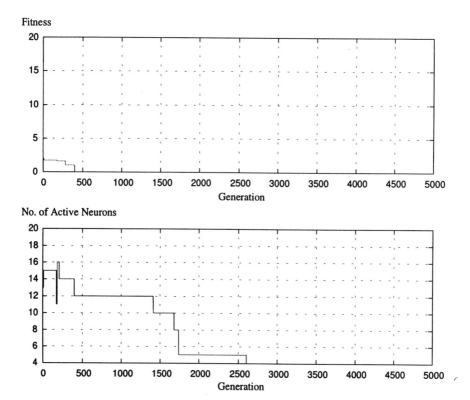

Fig. 7.16. Best chromosome vs generation

7.2 Fuzzy Logic

Ever since the very first introduction of the fundamental concept of fuzzy reasoning by Zadeh [245] in 1973, its use in engineering disciplines has been

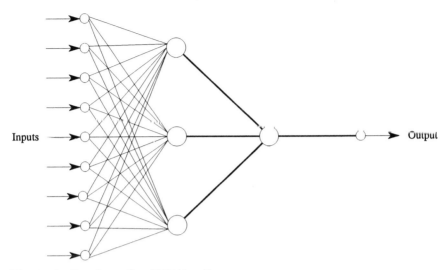

Fig. 7.17. Topology after 5000 iterations

widely studied. It has been reported that over 1,000 commercial and industrial fuzzy systems have been successfully developed in the space of last few years [154].

Its main attraction undoubtedly lies in the unique characteristics that fuzzy logic systems possess. They are capable of handling complex, nonlinear and sometimes mathematically intangible dynamic systems using simple solutions. Very often, fuzzy systems may provide a better performance than conventional non-fuzzy approaches with less development cost.

However, to obtain an optimal set of fuzzy membership functions and rules is not a easy task. It requires time, experience and skills of the operator for the tedious fuzzy tuning exercise. In principle, there is no general rule or method for the fuzzy logic set-up, although a heuristic and iterative procedure [169] for altering the membership functions to improve performance has been proposed, albeit that this is not optimal. Recently, many researchers have considered a number of intelligent schemes for the task of tuning the fuzzy set. The noticeable neural network approach [115] and the compatible GA methods [108, 119, 120, 165, 215] to optimize the membership functions and rules have become a trend for future fuzzy logic system development. It is our belief that the GA approach to optimize the fuzzy set is sound and that efforts in this direction should be continued.

Here, another innovative scheme is recommended. This approach differs from the other techniques in that it has the ability to reach an optimal set of memberships and rules without a known overall fuzzy set topology.

174   7. Hierarchical Genetic Algorithms in Computational Intelligence

This can be done only via the attributes of the HGA as discussed before. During the optimization phase, the membership functions need not be fixed. Throughout the genetic operations, a reduced fuzzy set including the number of memberships and rules will be generated. It is the purpose of this section to outline the essence of this technique based on fuzzy control.

7.2.1 Basic Formulation of Fuzzy Logic Controller

The fundamental framework of any fuzzy control system can be realized as shown in Fig. 7.18.

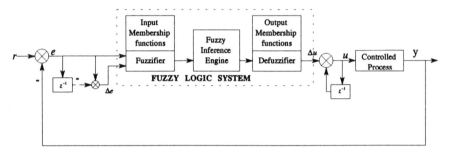

Fig. 7.18. Block diagram of genetic fuzzy logic controller

The operational procedure of the fuzzy logic controller (FLC) examines the receiving input variables e and Δe in a fuzzifying manner so that an appropriate actuating signal is derived to drive the system control input (u) in order to meet the ultimate goal of control. The favourable characteristic of the FLC lies is its ability to control the system without knowing exactly how the system dynamics behave. The degree of success using this method relies totally upon the judgment on the error signal e and Δe, where e is defined as $(e = r - y)$ and Δe is the digital rate of the change of e. The basic FLC arrangement is thus depicted in Fig. 7.19.

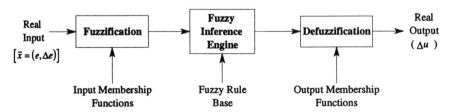

Fig. 7.19. Fuzzy logic system

7.2 Fuzzy Logic

Judging from the signals $(e, \Delta e)$, a mapping from $\underline{x} = (e, \Delta e) \in X \subset \Re^2$ to $\Delta u \in U \subset \Re$ can be performed. This process is generally known as *Fuzzification*. During the process, each input is classified into fuzzy subsets. Consider the error fuzzy set E as an example, this can be further divided into seven fuzzy subsets $(\mu_i^{(E)})$, defined as *Negative Large* $(NL,(\mu_1^{(E)}))$, *Negative Medium* $(NM, (\mu_2^{(E)}))$, *Negative Small* $(NS,(\mu_3^{(E)}))$, *Zero* $(ZE,(\mu_4^{(E)}))$, *Positive Small* $(PS,(\mu_5^{(E)}))$, *Positive Medium* $(PM,(\mu_6^{(E)}))$, *Positive Large* $(PM,(\mu_7^{(E)}))$. In general, these subsets can be constructed in the form of triangular membership functions as indicated in Fig. 7.20 in the dynamic range of $[e_{min}, e_{max}]$ as the minimum and maximum magnitude for signal e, respectively.

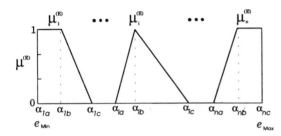

Fig. 7.20. Membership functions for fuzzy set E

The membership functions are defined as follows:

$$\mu_1^{(E)}(e) = \begin{cases} 1 & e \leq \alpha_{1b} \\ \frac{\alpha_{1c}-e}{\alpha_{1c}-\alpha_{1b}} & \alpha_{1b} < e < \alpha_{1c} \\ 0 & e \geq \alpha_{1c} \end{cases}$$

$$\mu_i^{(E)}(e) = \begin{cases} \frac{e-\alpha_{ia}}{\alpha_{ib}-\alpha_{ia}} & \alpha_{ia} < e \leq \alpha_{ib} \\ \frac{\alpha_{ic}-e}{\alpha_{ic}-\alpha_{ib}} & \alpha_{ib} < e < \alpha_{ic} \qquad \text{where } i = 2,\ldots,6 \\ 0 & e \leq \alpha_{ia} \text{ or } e \geq \alpha_{ic} \end{cases}$$

$$\mu_n^{(E)}(e) = \begin{cases} 0 & e \leq \alpha_{na} \\ \frac{e-\alpha_{na}}{\alpha_{nb}-\alpha_{na}} & \alpha_{nb} < e < \alpha_{na} \qquad \text{where } n = 7 \\ 0 & e \geq \alpha_{nb} \end{cases} \qquad (7.15)$$

Next, the degree of truth through the input membership functions is obtained and the same method applies to the membership functions for error rate fuzzy set (ΔE) and output fuzzy set (ΔU). Once the membership functions are installed, the fuzzy output (Δu) can be derived. This is commonly obtained by developing a set of fuzzy control rules which are capable of highlighting the concepts of fuzzy implication. The general form of the fuzzy rule is the "IF $\cdots$ and $\cdots$ THEN $\cdots$" structure.

Example 7.2.1. If (e is NS and Δe is PS) then Δu is ZE.
where NS and PS are the fuzzy subsets for e and Δe respectively; and ZE is the fuzzy subset for Δu.

Once the membership functions have been defined as indicated by Eqn. 7.15, (traditionally, this is normally done by hand and requires a good deal of experience), the appropriate fuzzy rules to govern the system can thus be developed. A typical fuzzy rule table is shown in Fig. 7.21.

e \ Δe	NL	NM	NS	ZE	PS	PM	PL
NL	NL	NL	NL	NM	NS	NS	ZE
NM	NL	NM	NM	NS	NS	ZE	PS
NS	NL	NM	NM	NS	ZE	PS	PS
ZE	NM	NM	NS	ZE	PS	PM	PM
PS	NS	NS	ZE	PS	PM	PM	PL
PM	NS	ZE	PS	PM	PM	PM	PL
PL	ZE	PS	PS	PM	PL	PL	PL

Fig. 7.21. IF-THEN rule

The rules $(R)_{i=1\to 7, j=1\to 7}$ for driving the system input Δu, as shown in Fig. 7.21, are then coded in the following manner:
$R_{1,1}$: If e is NL and Δe is NL then Δu is NL
$R_{1,2}$: If e is NL and Δe is NM then Δu is NL
$R_{1,3}$: If e is NL and Δe is NS then Δu is NL
$\vdots$
$R_{7,7}$: If e is PL and Δe is PL then Δu is PL
or
$$R = R_{1,1} \cup R_{1,2} \cup \cdots \cup R_{7,7} \tag{7.16}$$

Within these 49 individual rules, each rule is capable of governing the fuzzy output Δu. This then allows a wide variation for any of these rules to be valid at the same time. This phenomenon can be illustrated further by the following example.

Example 7.2.2. When a truth value of error signal (e) reaches a degree of truth to 0.7 on *NM* and 0.3 on *NS*; while the error rate signal Δe at the same time touches a degree of truth on 0.5 of *NS* and 0.9 of *ZE*, the associated governing rules can be obtained as indicated by the highlighted rules in Fig. 7.21. These rules are stated as
$R_{2,3}$: If e is NM and Δe is NS then Δu is NM
$R_{2,4}$: If e is NM and Δe is ZE then Δu is NS
$R_{3,3}$: If e is NS and Δe is NS then Δu is NM

$R_{3,4}$: If e is NS and Δe is ZE then Δu is NS

These can be directly mapped into the shaded output membership functions, NM and NS, for control, as shown in Fig. 7.22.

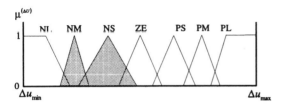

Fig. 7.22. Output membership functions

A union operation (Minimum Inferencing),

$$\mu_{(A \cap B)} = \min(\mu_A, \mu_B) \qquad (7.17)$$

where A and B are the fuzzy subsets may apply to determine the degree of truth on the output subsets of $R_{2,3}$, $R_{2,4}$, $R_{3,3}$ and $R_{3,4}$. The Minimum Inferencing on rule $R_{2,3}$ is used as a typical example to illustrate this point as shown in Fig. 7.23. The process should also be repeated for $R_{2,4}$, $R_{3,3}$ and $R_{3,4}$.

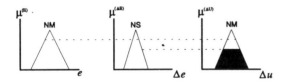

Fig. 7.23. Minimum inferencing on rule $R_{2,3}$

In this way, the degree of truth for the output fuzzy subsets can thus be obtained, and these are tabulated in Table 7.9 for the four rules $R_{2,3}$, $R_{2,4}$, $R_{3,3}$ and $R_{3,4}$.

Table 7.9. Example of minimum inferencing

	Error Fuzzy Subset	Error Rate Fuzzy Subset	Minimum Inferencing	Output Fuzzy Subset
$R_{2,3}$	NM (0.7)	NS (0.5)	min{0.7, 0.5}	NM (0.5)
$R_{2,4}$	NM (0.7)	ZE (0.9)	min{0.7, 0.9}	NS (0.7)
$R_{3,3}$	NS (0.3)	NS (0.5)	min{0.3, 0.5}	NM (0.3)
$R_{3,4}$	NS (0.3)	ZE (0.9)	min{0.3, 0.9}	NS (0.3)

178   7. Hierarchical Genetic Algorithms in Computational Intelligence

Here, it is clear that more than one degree of truth value can be assigned for an output fuzzy subset. These are, in this case, *NM* (0.3,0.5) or *NS* (0.7,0.3). To ensure a correct decision, a process of interaction (*Maximum composition*) may apply for the combination of NM and NS. This is indicated in Table 7.10.

Table 7.10. Example of maximum composition

	$R_{2,3}$	$R_{2,3}$	$R_{2,3}$	$R_{2,3}$	Max Composition	Degree of Truth
NM	0.5	-	0.3	-	max{0.5, 0.3}	0.5
NS	-	0.7	-	0.3	max{0.7, 0.3}	0.7

Finally, it was found that the degrees of truth for the output fuzzy subsets, *NM* and *NS* were 0.5 and 0.7, respectively. From this, the crisp value of the output variable Δu_o can be calculated via a process of *defuzzification* as shown in Fig. 7.24.

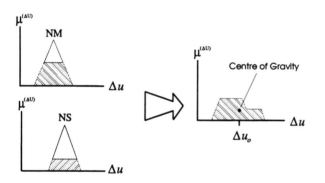

Fig. 7.24. Centre-of-gravity defuzzification

The actual value of Δu_o is calculated by the Centroid method:

$$\Delta u_o = \frac{\sum_i \Delta u_i \cdot \mu_i^{(\Delta U)}(\Delta u_i)}{\sum_i \mu_i^{(\Delta U)}(\Delta u_i)} \tag{7.18}$$

where $\mu_i^{(\Delta U)}$ is the membership function of fuzzy subset of ΔU.

This final value of Δu_o will be used to drive the system input (u) for the ultimate system control.

7.2.2 Hierarchical Structure

Having now learnt the complicated procedures of designing FLC, a practical realization of this system is not easy to determine. The dynamic variation of fuzzy input membership functions and the interpretation of governing rules for fuzzy output are the main stumbling blocks to this design. Manually operating procedures for these variables might not only yield a sub-optimal performance, but could also be dangerous if the complete fuzzy sets were wrongly augmented.

Considering that the main attribute of the HGA is its ability to solve the topological structure of an unknown system, then the problem of determining the fuzzy membership functions and rules could also fall into this category. This approach has a number of advantages:

- an optimal and the least number of membership functions and rules are obtained;
- no pre-fixed fuzzy structure is necessary; and
- simpler implementing procedures and less cost are involved.

Hence, it is the purpose of this section to introduce the HGA for the designing of FLC [220]. The conceptual idea is to have an automatic and intelligent scheme to tune the fuzzy membership functions and rules, in which the closed loop fuzzy control strategy remains unchanged, as indicated in Fig. 7.25.

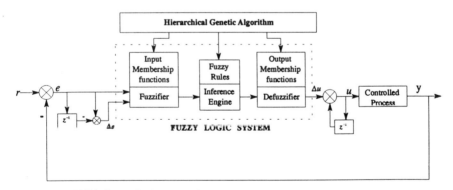

Fig. 7.25. HGA fuzzy logic control system

Chromosome of HGA. Similar to other uses of the HGA, the hierarchical chromosome for the FLC structure must be correctly formulated. In this case, the chromosome of a particular fuzzy set is shown in Fig. 7.26. The chromosome consists of the usual two types of genes, the control genes and parameter genes. The control genes, in the form of bits, determine the membership function activation, whereas the parameter genes (similar to those stated in Eqn. 7.15) are in the form of real numbers to represent the membership functions.

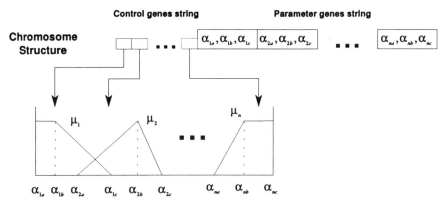

Fig. 7.26. Hierarchical membership chromosome structure

With the two input fuzzy sets of error signals, (e), error rate (Δe) and the output fuzzy set of Δu, we can thus construct the overall membership chromosome structure as in Fig. 7.27.

Fig. 7.27. HGA chromosome structure

The parameter genes (z_p) of the membership chromosome take the form:

$$z_p = \{\alpha_{1a}^{(E)}, \alpha_{1b}^{(E)}, \alpha_{1c}^{(E)}, \ldots, \alpha_{ma}^{(E)}, \alpha_{mb}^{(E)}, \alpha_{mc}^{(E)}, \beta_{1a}^{(\Delta E)}, \beta_{1b}^{(\Delta E)}, \beta_{1c}^{(\Delta E)}, \ldots,$$
$$\beta_{na}^{(\Delta E)}, \beta_{nb}^{(\Delta E)}, \beta_{nc}^{(\Delta E)}, \gamma_{1a}^{(\Delta U)}, \gamma_{1b}^{(\Delta U)}, \gamma_{1c}^{(\Delta U)} \ldots \gamma_{pa}^{(\Delta U)}, \gamma_{pb}^{(\Delta U)}, \gamma_{pc}^{(\Delta U)}\}$$

where m, n and p are the maximum allowable number of fuzzy subset of E, ΔE and ΔU, respectively; $\alpha_{ia}^{(E)}, \alpha_{ib}^{(E)}, \alpha_{ic}^{(E)}$ define the input membership function of i-th fuzzy subset of E; $\beta_{ja}^{(\Delta E)}, \beta_{jb}^{(\Delta E)}, \beta_{jc}^{(\Delta E)}$ define the input membership function of j-th fuzzy subset of ΔE; and $\gamma_{ka}^{(\Delta U)}, \gamma_{kb}^{(\Delta U)}, \gamma_{kc}^{(\Delta U)}$ define the output membership function of k-th fuzzy subset of ΔU.

To obtain a complete design for the fuzzy control design, an appropriate set of fuzzy rules is required to ensure system performance. At this point, it should be stressed that the introduction of the control genes is done to govern the number of fuzzy subsets E, ΔE and ΔU. As a result, it becomes impossible to set a universal rule table similar to the classic FLC approach for each individual chromosome. The reason for this is that the chromosomes may vary from one form to another, and that each chromosome also has a different number of fuzzy subsets.

Therefore, the fuzzy rules based on the chromosome set-up should be classified. For a particular chromosome, there should be w, x and y active subsets with respect to E, ΔE and ΔU in an independent manner. This can be represented by a rule table, as shown in Table 7.11, with a dimension $w \times x$. Then, each cell defines a rule for the controller, i.e. the $i-j$ element implies rule $R_{i,j}$:

$R_{i,j}$: If e is E_i and Δe is D_j then Δu is U_k

where E_i, D_j, U_k are now the linguistic name, similar to "Large", "Small" and so on, to characterize the fuzzy subsets of error, error rate and output set, respectively.

Table 7.11. The rule base in tabular form

	D_1	D_2	$\cdots$	D_j	$\cdots$	D_x
E_1	U_1	U_2		$\cdots$		U_j
E_2	U_2	U_3		$\cdots$		U_j
$\vdots$			$\ddots$	$\vdots$		
E_i		$\cdots$		U_k	$\cdots$	
$\vdots$				$\vdots$		
E_w	U_i			$\cdots$		U_y

The Fuzzy Rule Chromosome, $H_{(w,x,y)}$, is then formulated in the form of an integer matrix where

$$H_{(w,x,y)} = \{h_{i,j} : h_{i,j} \in [1,y] \quad \forall i \leq w,\ j \leq x\} \tag{7.19}$$

Example 7.2.3. For a Fuzzy Rule Chromosome with $w=2$, $x=2$ and $y=2$, then,

$$H_{(2,2,2)} = \begin{bmatrix} 1 & 2 \\ 2 & 2 \end{bmatrix} \tag{7.20}$$

From $H_{(2,2,2)}$, there are four rules:
$R_{1,1}$: If e is E_1 and Δe is D_1 then Δu is U_1
$R_{1,2}$: If e is E_1 and Δe is D_2 then Δu is U_2
$R_{2,1}$: If e is E_2 and Δe is D_1 then Δu is U_2
$R_{2,2}$: If e is E_2 and Δe is D_2 then Δu is U_2

7. Hierarchical Genetic Algorithms in Computational Intelligence

Genetic Cycle. Once the formulation of the chromosome has been set for the fuzzy membership functions and rules, genetic operation cycle can be performed. This cycle of operation for the FLC optimization using an HGA is illustrated in Fig. 7.28.

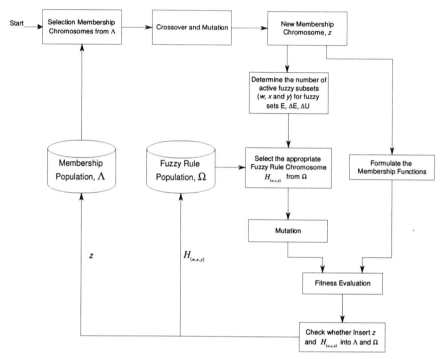

Fig. 7.28. Genetic cycle for fuzzy logic system optimization

Population. There are two population pools, (Λ) and (Ω), for storing the membership and fuzzy rule chromosomes, respectively. The HGA chromosomes are grouped in Λ, while the fuzzy rule chromosomes are stored in the fuzzy rule population, Ω. Within Ω, there should be a total number of $(m-1) \times (n-1) \times (p-1)$ fuzzy rule sets. However, only one particular single rule set can be matched with $H_{(w,x,y)}$ in order to satisfy the chromosome that possesses the w, x and y active fuzzy subsets of E, ΔE and ΔU, respectively.

Genetic Operations. Considering that there are various types of gene structure, a number of different genetic operations have been designed. For the crossover operation, a one point crossover is applied separately for both the control and parameter genes of the membership chromosomes within certain operation rates. There is no crossover operation for fuzzy rule chromosomes since only one suitable rule set $H_{(w,x,y)}$ can be assisted.

Bit mutation is applied for the control genes of the membership chromosome. Each bit of the control gene is flipped ("1" or "0") if a probability test is satisfied (a randomly generated number, r_c, is smaller than a pre-defined rate). As for the parameter genes, which are real-number represented, random mutation is applied. A special mutation operation has been designed to find the optimal fuzzy rule set. This is a delta shift operation which alters each element in the fuzzy rule chromosome as follows:

$$h_{i,j} = h_{i+\Delta i, j+\Delta j} \qquad (7.21)$$

where Δi, Δj have equal chance to be 1 or -1 with a probability of 0.01.

Fitness Evaluation. Before evaluating the fitness value of the chromosome pair (z, H), their phenotype must be obtained. In some cases, a direct decoding of the membership chromosome may result in invalid membership functions. For example, Fig. 7.29 represents an invalid membership function for error fuzzy set because the ranges $(\alpha_{1c}, \alpha_{3a})$ and $(\alpha_{4c}, \alpha_{7c})$ are unclassified (only the error set is shown for clarity).

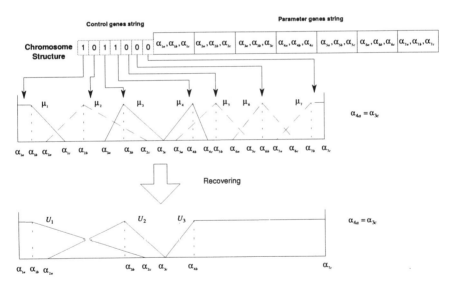

Fig. 7.29. Recovery of invalid fuzzy membership functions

To ensure that there was no undefined region, a remedial procedure was operated to ensure validation. The decoded fuzzy membership functions were recovered as shown by the final membership characteristics in Fig. 7.29. It should be noted that the parameter gene remained unaltered but merely changed the interpretation of its form. In this way, the complexity of tuning the fuzzy memberships and rules can thus be optimized and the overall structure can be greatly reduced.

Together with the fuzzy rule table generated from the fuzzy rule chromosome, a full set of FLC can then be designed. A fitness value, $f(z, H)$, can then be determined which reflects the fitness of the FLC.

Insertion Strategy. The grouping method, as described in Chap. 4, was adopted in order to find optimal membership functions of E, ΔE and ΔU and the appropriate fuzzy rules. The population of membership chromosomes Λ is divided into several subgroups, $S_{(i,j,k)}$, such that

$$\Lambda = S_{(2,2,2)} \cup S_{(2,2,3)} \ldots \cup S_{(2,n,p)} \cup S_{(3,2,2)} \ldots \cup S_{(m,n,p)} \quad (7.22)$$

and

$$S_{(i,j,k)} \cap S_{(w,x,y)} \neq \emptyset \quad \forall (i \neq w \vee j \neq x \vee k \neq y) \quad (7.23)$$

where $S_{(i,j,k)}$ is the subgroup of chromosome that represents those with i, j and k active fuzzy subsets for E, ΔE and ΔU, respectively.

The maximum number of subgroups in Λ is thus $(m-1) \times (n-1) \times (p-1)$. A concurrent factor, λ (typically assigned as 3-5), is used to define the maximum elements stored in the membership subgroup,

$$size\left[S_{(i,j,k)}\right] \leq \lambda \quad \forall\, 2 \leq i \leq m,\ 2 \leq j \leq n, 2 \leq k \leq p \quad (7.24)$$

where $size\left[S_{(i,j,k)}\right]$ is the number of elements in $S_{(i,j,k)}$.

Table 7.12 explains the insertion strategy for new membership chromosome (z) with active w, x, and y fuzzy subsets for sets E, ΔE and ΔU, respectively, with new fuzzy rule chromosome, $H_{(w,x,y)}$.

The complete genetic cycle continues until some termination criteria, for example, meeting the design specification or number of generation reaching a predefined value, are fulfilled.

7.2.3 Application I: Water Pump System

To test the design of the HGA fuzzy logic controller, an experimental piece of equipment which consisted of a water pump having a 1.5 horse power engine and a water tank was used for the investigation. It simulated the constant pressure booster pump system designed for a water supply system [102].

The compensated actuator unit was the water pump with a variable frequency converter (VFC) attached. The actuating signal came from a pressure sensor placed in a pipe downstream and its output signal was fed back into the VFC to change the pump speed. A schematic diagram of the water supply system is shown in Fig. 7.30.

7.2 Fuzzy Logic

Table 7.12. Insertion strategy

At generation $(k+1)$

Step 1:
If $\left\{ S_{(w,x,y)} = \emptyset \quad \text{or} \quad size\left[S_{(w,x,y)}\right] < \lambda \right\}$
then
$\quad S_{(w,x,y)} = S_{(w,x,y)} \cup \{z\} \quad$ and
$\quad \Omega = \left\{ H_{(i,j,k)} : H_{(i,j,k)} \in \Omega \quad i \neq w, j \neq x, k \neq y \right\} \cup \left\{ H_{(w,x,y)} \right\}$
else
$\quad$ goto step 2

Step 2:
If $\left\{ f(z, H_{(w,x,y)}) < f_{max} = \max\left\{ f(z_i, H_{(w,x,y)}), \, \forall z_i \in S_{(w,x,y)} \right\} \right\}$
then
$\quad S_{(w,x,y)} = \left\{ z_i : F(z_i) < f_{max}, \, z_i \in S_{(w,x,y)} \right\} \cup \{z\} \quad$ and
$\quad \Omega = \left\{ H_{(i,j,k)} : H_{(i,j,k)} \in \Omega \quad i \neq w, j \neq x, k \neq y \right\} \cup \left\{ H_{(w,x,y)} \right\}$
else
$\quad$ goto step 3

Step 3:
$\quad$ Exit

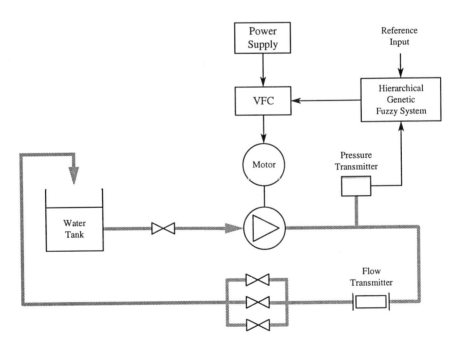

Fig. 7.30. Experimental apparatus

The underlying problem of the control is the nonlinear characteristic for both QH characteristic of the actuating power source (fan speed) and the dissipated energy due to the pipe characteristics, as shown in Fig. 7.31. When the control loop is closed, the dissipated energy of the pipe is to be suppressed adequately by the actuating power so as to keep the variation of the perturbed pressure at a minimum level, and yet the water flow rate is allowed to changed freely to suit the demand.

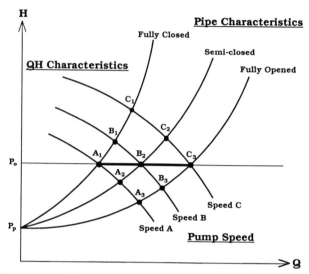

Fig. 7.31. Pressure and flow rate characteristics

Fuzzy Learning and Optimization. The parameter setting of HGA to optimize the required fuzzy subsets is tabulated in Table 7.13 with a command reference signal in the following form:

$$r(k) = \begin{cases} 1000 & 1 \le k \le 400 \\ 1200 & 401 \le k \le 800 \\ 1000 & 801 \le k \le 1200 \end{cases} \quad (7.25)$$

where k is the sample.

The design specification is to meet two objective functions:
1. Minimum output steady state error

$$f_1 = \frac{1}{100} \left[\sum_{k=701}^{800} (y(k) - r(k))^2 + \sum_{k=1101}^{1200} (y(k) - r(k))^2 \right] \quad (7.26)$$

where $r(k)$ and $y(k)$ are the reference and plant outputs, respectively; and

7.2 Fuzzy Logic

Table 7.13. Parameters of HGA for fuzzy controller

	Membership Chromosome		Fuzzy Rule Chromosome
	Control Genes	Connection Genes	
Representation	Binary	Real Number	Integer
Population Size No. of Offspring		20 2	216 (m=n=p=7) –
Crossover Crossover Rate	One-Point 1.0	One-Point 1.0	– –
Mutation Mutation Rate	Bit Mutation 0.02	Random Mutation 0.02	Eqn. 7.21 Eqn. 7.21
Selection	Roulette Wheel Selection on Rank		Based on the no. of active fuzzy subsets
Reinsertion	Table 7.12		Direct replacement

2. Minimum overshoot and undershoot

$$f_2 = p_1 + p_2 \tag{7.27}$$

where

$$\text{overshoot:} \quad p_1 = \begin{cases} \frac{y_{max} - r(401)}{r(401) - r(400)} & y_{max} > r(401) \\ 0 & y_{max} \leq r(401) \end{cases}$$

and

$$\text{undershoot:} \quad p_2 = \begin{cases} \frac{r(801) - y_{min}}{r(800) - r(801)} & y_{min} < r(801) \\ 0 & y_{min} \geq r(801) \end{cases}$$

with $y_{max} = max\{y(k) : 401 \leq k \leq 800\}$ and $y_{min} = min\{y(k) : 801 \leq k \leq 1200\}$

Based on the design procedures that have already been described in the above sections, the obtained plant output response is obtained after 20 generations of computation. The result is shown in Fig. 7.32.

The same set of procedures is repeated again, in this case, the maximum allowable overshoot and undershoot are now set to 0%. This result is depicted in Fig. 7.33. These clearly indicate that the HGA fuzzy control design scheme is fully justified. The corresponding FLC design parameters as well as the obtained fuzzy subsets and membership functions (Figs. 7.34, 7.35 and 7.36) are tabulated as follows:

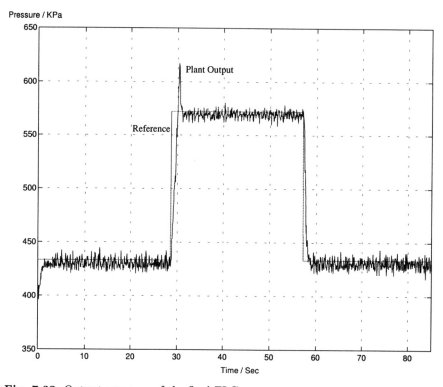

Fig. 7.32. Output response of the final FLC

Table 7.14. Optimal membership functions obtained

Control Genes	Parameter Genes	Fuzzy Subset
$z_c^{(E)} = [0010110]$	$z_p^{(E)} = [(-100, -98, -48), (-97, -47, -45),$ $(-46, -42, -29), (-41, -22, -13),$ $(-19, 4, 15), (4, 17, 40), (29, 47, 100)]$	Fig. 7.34
$z_c^{(\Delta E)} = [1000101]$	$z_p^{(\Delta E)} = [(-100, -90, -66), (-80, -59, -46),$ $(-51, -18, -38), (19, 50, 67),$ $(59, 79, 79), (79, 80, 94), (80, 97, 100)]$	Fig. 7.35
$z_c^{(\Delta U)} = [1001010]$	$z_p^{(\Delta U)} = [(-100, -97, -95), (-96, -75, -65),$ $(-66, -64, 4), (-52, 8, 25),$ $(12, 31, 31), (31, 32, 59), (56, 92, 100)]$	Fig. 7.36

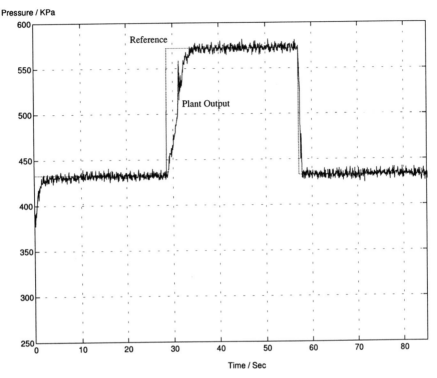

Fig. 7.33. Output response of the best FLC with multiobjective

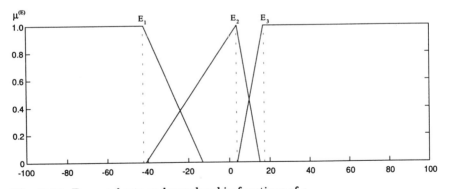

Fig. 7.34. Fuzzy subsets and membership functions of e

190   7. Hierarchical Genetic Algorithms in Computational Intelligence

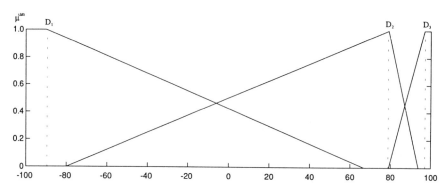

Fig. 7.35. Fuzzy subsets and membership functions of Δe

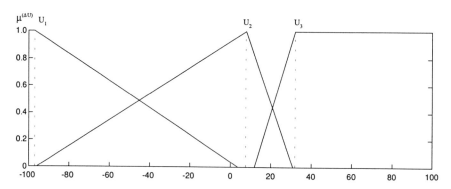

Fig. 7.36. Fuzzy subsets and membership functions of Δu

- The membership chromosome is tabulated in Table 7.14 with the associated figures of each fuzzy set.
- The fuzzy rule chromosome is obtained as

$$H = \begin{bmatrix} 0 & 1 & 1 \\ 1 & 1 & 2 \\ 1 & 2 & 2 \end{bmatrix}$$

which implies a rule table as in Table 7.15.

Table 7.15. Optimal rule table

		Error Rate Fuzzy Set		
		D_1	D_2	D_3
Error	E_1	U_1	U_2	U_2
Fuzzy	E_2	U_2	U_2	U_3
Set	E_3	U_2	U_3	U_3

It has been shown that a reduced size of subsets of fuzzy membership functions and rules is obtained by the use of HGA while still meeting the requirements of system performance. This result is considered to be compatible to those obtained using the conventional fuzzy logic design schemes.

Closed Loop Performance. To further verify the performance of the obtained FLC design, two sets of experimental tests were conducted. Firstly, an irregular square waveform command signal was applied to the water supply system at a number of operating points. It can be seen from Fig. 7.37 that the pressure rise and fall characteristics closely track the set points.

Secondly, a nominal operating point of the water pipe pressure was chosen for dynamic disturbance testing. The disturbance was created by turning the water valve along the pipe to the "on" and "off" state while the water supply was in operation. Fig. 7.38 shows that the pressure recovered well from the disturbance despite the condition of the water valve.

7.2.4 Application II: Solar Plant

The FLC controller design is applied for a distributed collector field of the solar plant in Almería (Spain). The collector field of the solar plant in Almería consists of 480 distributed solar ACUREX collectors. These collectors are arranged in 20 rows forming 10 parallel loops. Each loop is about 172 metres long. The collector uses a parabolic mirror to reflect solar radiation onto a pipe where the oil is heated. The energy collected is transferred to a storage

192   7. Hierarchical Genetic Algorithms in Computational Intelligence

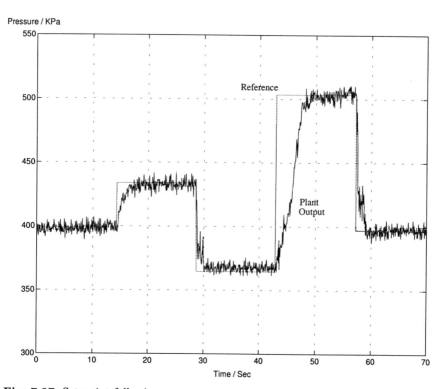

Fig. 7.37. Set-point following

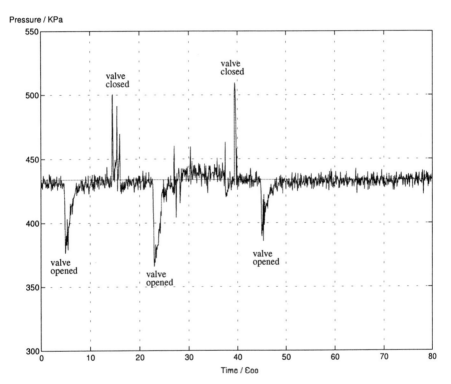

Fig. 7.38. Closed loop disturbance

tank and then to either a steam generator for electrical power generation or the heater exchanger for a desalination plant. The schematic diagram of the collector field is shown as Fig. 7.39.

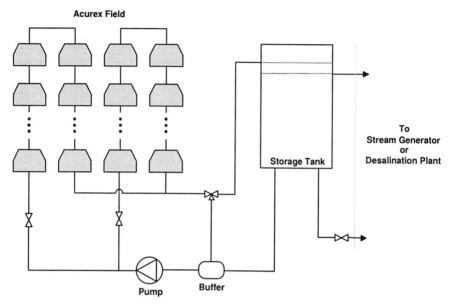

Fig. 7.39. Schematic diagram of the ACUREX distributed solar collector field

The objective of the control system is to maintain the outlet oil temperature at a desired reference by adjusting the flow of the oil in spite of disturbances, such as the changes in the solar radiation level, mirror reflectivity or inlet oil temperature. Since the disturbances leads to significant variations in the dynamic characteristics of the system, it would be difficulties in obtaining satisfactory performance with a fixed parameter controller over a wide operating range.

The fuzzy logic controller in Fig. 7.40 is applied to the solar power plant. An additional feedforward term is included after the FLC on a steady state energy balance, which makes an adjustment in the flow input, aimed at eliminating the change in outlet temperature as a result of the variations in solar radiation (I) and inlet temperature (T_{in}). The complete control scheme is depicted in Fig. 7.40.

The proposed control scheme was applied to the simulated plant. A testing data simulating the working condition such as the solar radiation and inlet oil temperature etc. of the plant on September 16, 1991 was used. The plant response is shown in Fig. 7.41. From the result, it can be seen that the

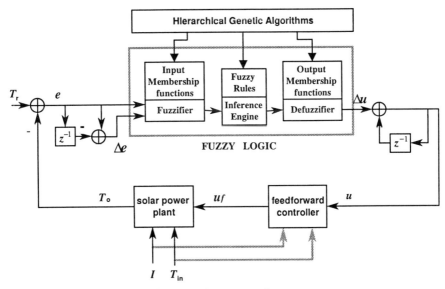

Fig. 7.40. Control scheme for the solar power plant

temperature of the outlet oil is well tracking the set point of the desired oil temperature, with a small overshoot and a fast rising time, after the starting phase.

The final membership functions and the corresponding rule table are shown in Fig. 7.42 and Table 7.16, respectively.

This shows that a reduced size of subsets ($3 \times 5 \times 5$) of fuzzy membership functions and rules is obtained by the use of HGA, while the requirements of system performance have not been deteriorated. This result is shown in Fig. 7.41 and considered to be compatible to those obtained using the conventional fuzzy logic design schemes [182].

Table 7.16. Optimal rule table

		Error Rate Fuzzy Set				
		D_1	D_2	D_3	D_4	D_5
Error	E_1	U_1	U_2	U_2	U_3	U_3
Fuzzy	E_2	U_2	U_3	U_3	U_3	U_4
Set	E_3	U_3	U_3	U_4	U_5	U_5

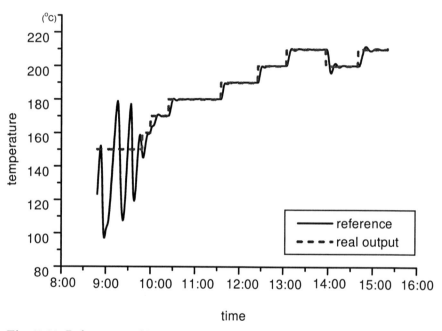

Fig. 7.41. Reference tracking

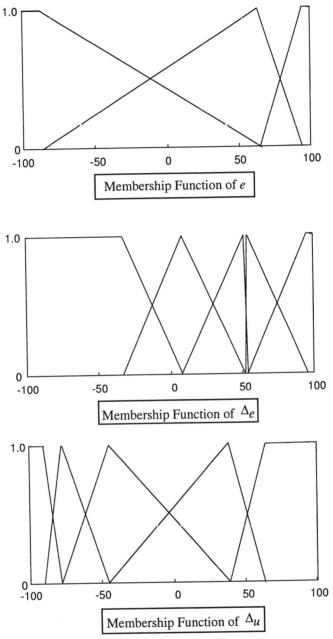

Fig. 7.42. Final fuzzy subsets and membership functions for solar power plant

8. Genetic Algorithms in Speech Recognition Systems

In speech recognition, the training (or learning) process plays an important role. When a good training model for a speech pattern is obtained, this not only enhances the speed of recognition tremendously, but also improves the quality of the overall performance in recognizing the speech utterance. In general, there are two classic approaches for this development, namely, Dynamic Time Warping (DTW) and the Hidden Markov Model (HMM).

In this chapter, GA is applied to solve involved nonlinear, discrete and constrained problems for the DTW, while for the case of HMM, a unique approach to obtain a HMM model using parallel GA architecture is described. Because of the intrinsic properties of GA, the associated non-trival K-best paths of DTW can be identified without extra computational cost, and the concept of parallel GA in HMM modelling is recognized as being a sound way for the discovery of the best, if not optimal template reference model.

8.1 Background of Speech Recognition Systems

The dream of speaking to computers only have come true due to the technological advances that have been made in the areas of speech processing and pattern recognition. These technologies not only allowed us to model speech utterances in a correct manner but also to classify the label of each particular utterance. The task of speech recognition essentially is to capture a speech signal from an input device such as microphone and to identify the label of this signal.

However, this task is simple, because of the time-varying properties of speech signals. For example, if one person speaks the same word for several times, the utterance will have different characteristics. When the Chinese word 'one' was spoken four times, it can be seen from Fig. 8.1 that the speech utterances did not have the same outlook across the entire utterance range. Thus, in the course of developing a speech recognition system, the crux of matter would seem to be to perform a robust and accurate matching in the recognition process. This is by no means an easy task, as speech utterance in

200   8. Genetic Algorithms in Speech Recognition Systems

general is fuzzy, varies in time and is sometimes unpredictable.

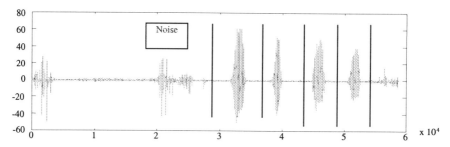

Fig. 8.1. Four different outlook for the same utterance of Chinese word 'one'

With many years of hard and good work undertaken by the researchers in this field, fortunately, many speech recognition systems have been successfully launched. Examples of such applications are the automation of operator services by AT&T and Northern Telecom, the automation of directory services by NYNEX and Northern Telecom, and the stock quotation services by Bell Northern Research [181]. In this chapter, we do not attempt to cover the complete history and technical background in speech recognition. Instead, we focus on the application of the Genetic Algorithm to speech recognition systems and how this can improve the performances of the system especially in the area of speech pattern matching.

8.2 Block Diagram of a Speech Recognition System

A general block diagram of a speech recognition system is shown in Fig. 8.2.

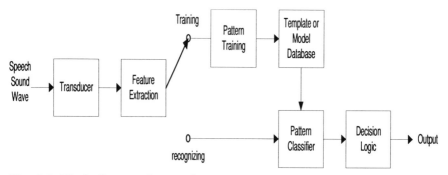

Fig. 8.2. Block diagram of a speech recognition system

In a speech recognition system, speech utterances must be converted to digital representations for computer processing. Therefore, speech signals or sound waves will first pass through a transducer such as microphone and be digitized into a set of discrete data points. Of course, this conversion must be accurate enough to represent the original signal and ensure that the integrity of the speech signals is preserved.

According to Shannon's sampling theorem [195], the originality of the signal will not be distorted if we sample signals at the Nyquist rate [160] that is equal to two times the bandwidth of the original signal. It is well known that human speech in general lies below 4 KHz, thus, a sample rate of 8 KHz is sufficient to reconstruct the original signal without loss of originality. Then, an analog-to-digital converter with 12 bits resolution and 16,000 Hz of sampling frequency can be used for the conversion of the analog speech signal into its digital form.

Once the raw speech signals are digitized, the set of discrete points is further converted to feature vectors so that the speech signal can be best represented with a minimum number of points. In fact, the purpose of feature analysis is also to distill the necessary information from the raw speech signals [181]. The feature vectors can be obtained either in time domain or frequency domain. In time domain, the feature vectors provide useful representation such as energy, pitch, zero-crossings [170]. However, these measurements do not reveal the entire picture of speech signals completely. These measurements sometimes can only be used to detect the difference between signals and noise for end-point detection or the voiced and non-voiced parts of speech signals. Whereas in frequency domain, spectral analysis can be used to capture the envelope characteristics of the signal.

Spectral analysis methods can be applied to speech signals over a short period of time ranging from 10ms to 30ms in order to extract the short time spectral envelope characteristics of the signal. There are many techniques which can be used for spectral analysis such as Discrete Fourier transform (DFT) [170], Fast Fourier transform (FFT) [38, 99], filter bank [40, 41], linear prediction [144] and ceptral analysis [163]. Among these techniques, Atal [6] has pointed out that ceptral analysis has been found to be one of the most effective feature vectors that can provide the best results. One example is the HTK recognizer developed in Cambridge University [242] in which researchers used the 12 mel-frequency cepstral coefficients (MFCC) [243] and the signal energy to form the basic 13-element acoustic feature vector. In addition, they also demonstrated empirically that the MFCC can improve the system performance of most speech recognition systems [180].

Once the process of extracting feature vector has completed, two other modes of operation have to be performed in speech recognition systems, namely (1) training and (2) recognition. These operations ensure that the speech utterance are labeled and stored in a database for the purpose of recognition.

As each utterance has its own distinct characteristics in terms of energy or spectral features, it is necessary to provide a way to find a single or multiple utterance(s) to be best represented for all the members of the same class. This is known as a labeling process and commonly called "Training" or "learning" in a speech recognition system. To continue on from our previous example for the recognition of the Chinese word 'one', this particular utterance has to be trained and then stored in the database. In this case, the same word 'one' with four similar but varying sets of feature vectors can be used for training.

It is also natural to believe that the more utterances used for training, the better accuracy can be achieved. However, there is a limit. In reality, it is not possible to collect all the utterances for training as the speech utterances do not have consistent characteristics at all times. However, error will creep into the recognition process if there is lack of sufficient trained utterances.

In the recognition mode, the unknown utterance is captured and converted into a set of feature vectors. The mission of the recognition system is to compare this set of feature vectors with each reference pattern in the database by using some kind of pattern classification technique. Depending upon the technique used, a measure of similarity between the feature pattern and each reference pattern is computed. The one with the largest value of similarity would be considered as the most likely candidate of the unknown utterance. The "Pattern Classifier" indicated in Fig. 8.2 illustrates this identification task. In general, pattern classifier involves two major techniques: the DTW [187, 112, 171, 156] and the HMM [172].

The DTW algorithm is a template matching method which computes the minimum distortion between two patterns. If the two patterns are found to be similar to each other, then the difference or distortion measurement should be small. On the other hand, a large value implies dissimilarity between the two patterns.

This type of template matching method has the advantage that the templates of spectral sequences can be easily constructed using a multiple-speaker database, and only the distance relationship is used for the calculation of the difference between the trained utterances [69]. This technique is now well developed and provides a good recognition method in a number of practical applications [24, 68, 27].

HMM is a statistical method using probability. It can be used to represent a speech utterance sequence for training, and also generates the sequence for the process of recognition. The measure in term of maximum likelihood probability is computed and the larger the value of measure, the higher is the chance of similarity between the two patterns being reached. It has been shown that HMM based speech recognizers can provide better and more robust results than those obtained from the template matching method. The fundamentals of both DTW and HMM techniques can be found in [172, 173]. The purpose of the following sections is to outline the main feature of GA applications for speech recognition based on the DTW and HMM methodologies.

8.3 Dynamic Time Warping

As DTW is a technique commonly used to evaluate the similarity between two different utterances based on the measure of local minimum distance, there is a need to normalize the speaking rate of fluctuation in order that the utterance comparison can be as meaningful as possible before a recognition decision is made. The technique of dynamic programming is one of the most common methods applied for solving the nonlinear DTW algorithm [156, 173].

Let $\mathbf{X} = (\mathbf{x}_1, \mathbf{x}_2, \cdots, \mathbf{x}_N)$ and $\mathbf{Y} = (\mathbf{y}_1, \mathbf{y}_2, \cdots, \mathbf{y}_M)$ represent two speech patterns, where $\mathbf{x}_i$ and $\mathbf{y}_j$ are the short-time acoustic feature vectors such as the linear predictive coefficients, ceptrum coefficients, pitch, energy or any set of acoustic features, etc. Let i_x and i_y be the time indices of $\mathbf{X}$ and $\mathbf{Y}$, respectively. The dissimilarity measure between $\mathbf{X}$ and $\mathbf{Y}$ is defined as the short-time spectral distortions $d(\mathbf{x}_{i_x}, \mathbf{y}_{i_y})$ of the two speech patterns which are denoted as $d(i_x, i_y)$, where $i_x = 1, 2, \ldots, N$ and $i_y = 1, 2, \ldots, M$.

A general time alignment and normalization scheme involves the use of two warping functions, ϕ_x and ϕ_y, which relate the indices of the two speech patterns, i_x and i_y, respectively for a common "normal" time axis k, i.e.,

$$i_x = \phi_x(k)$$
$$i_y = \phi_y(k) \quad \text{where} \quad k = 1, 2, \ldots, T$$

A global pattern dissimilarity measure $d_\phi(\mathbf{X}, \mathbf{Y})$ is defined based upon the warping function pair $\phi = (\phi_x, \phi_y)$ as the accumulated distortion over the entire utterance. This can be stated as follows:

8. Genetic Algorithms in Speech Recognition Systems

$$d_\phi(\mathbf{X}, \mathbf{Y}) = \sum_{k=1}^{T} \frac{d(\phi_x(k), \phi_y(k)) \, m(k)}{M_\phi} \tag{8.1}$$

where $d(\phi_x(k), \phi_y(k))$ is a short-time spectral distortion defined for the $\mathbf{x}_{\phi_x}(k)$ and $\mathbf{x}_{\phi_y}(k)$; $m(k)$ is a non-negative (path) weighting coefficient and M_ϕ is a (path) normalizing factor. Obviously, there are an extremely large number of warping function pairs that satisfy the above requirement. Then the next issue is to determine a distortion measurement for all the possible paths. One natural and popular choice is to define the dissimilarity function $d(\mathbf{X}, \mathbf{Y})$ as the minimum of $d_\phi(\mathbf{X}, \mathbf{Y})$, over all possible paths, such that

$$d(\mathbf{X}, \mathbf{Y}) = \min_\phi d_\phi(\mathbf{X}, \mathbf{Y}) \tag{8.2}$$

where ϕ must satisfy a number of constraints that reflects the reality of speech signals, such as endpoint constraints, monotonic conditions, local continuity constraints, slope constraints, and allowable regions, etc [173]. These conditions are discussed briefly below:

Endpoint constraints — The uttered signal in the signal stream must have well-defined endpoints that mark the beginning and ending frames of the signal for recognition, i.e.,

$$\begin{aligned} \text{Beginning point:} \quad & \phi_x(1) = 1, \quad \phi_y(1) = 1 \\ \text{Ending point:} \quad & \phi_x(T) = N, \quad \phi_y = (T) = M \end{aligned} \tag{8.3}$$

It should be noted that the endpoints are easily disrupted by background noise. As a result, this can result in inaccurate endpoint estimation. Hence, rather than rely on the perfection of the endpoint detection mechanism, instead, the endpoint constraints should be relaxed as follows:

$$\begin{aligned} \text{Beginning point:} \quad & \phi_x(1) = 1 + \Delta x, \quad \phi_y(1) = 1 + \Delta y \\ \text{Ending point:} \quad & \phi_x(T) = N - \Delta x, \quad \phi_y(T) = M - \Delta y \end{aligned} \tag{8.4}$$

Monotonicity — The temporal order is a special feature of acoustic information in speech patterns. Therefore, the warping path cannot be traversed in reverse. A local monotonic constraint should be imposed to prevent such an occurrence. This can be done by restricting ϕ, i.e.,

$$\begin{aligned} \phi_x(k+1) &\geq \phi_x(k) \\ \phi_y(k+1) &\geq \phi_y(k) \end{aligned} \tag{8.5}$$

8.3 Dynamic Time Warping

Local Continuity — This property ensures the proper time alignment and maintains the integrity of the acoustic information. The local continuity constraint imposed on ϕ is as follows:

$$\phi_x(k+1) - \phi_x(k) \leq 1$$
$$\phi_y(k+1) - \phi_y(k) \leq 1 \quad (8.6)$$

Slope Weighting — Neither too steep nor too gentle a gradient should be allowed in ϕ, otherwise unrealistic time-axis warping may appear. A too steep or too gentle gradient in ϕ implies comparison of a very short pattern with a relatively long one. This comparison is generally not realistic since the two patterns do not possess these characteristics. Therefore, the weighting function $m(k)$ in Eqn. 8.1 controls the contribution of the short-time distortion $d(\phi_x(k), \phi_y(k))$ which is used to eliminate the undesired time-axis warping.

Slope constraint is regarded as a restriction on the relation of several consecutive points on the warping function rather than a single point. As indicated in Fig. 8.3(a), if ϕ moves along the horizontal for m times, then ϕ must move along in a diagonal direction for at least n times before stepping against the horizontal direction. The effective intensity of the slope constraint can be measured by the parameter $P = \frac{n}{m}$. H. Sakoe and S. Chiba [188] suggested four types of slope constraints with different values of P. These can be seen in Fig. 8.3(c),

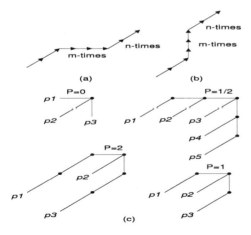

Fig. 8.3. Slope constraints on warping path. (a) Minimum slope. (b) Maximum slope. (c) Four types of slope constraints suggested by H. Sakoe and S.Chiba

Allowable regions — Because of the slope constraints, certain portions of (i_x, i_y) are excluded from the searching region. The allowable region can be defined by two parameters: Q_{max} and Q_{min} the maximum slope and minimum slope of searching steps in the DP-searching path, respectively. For example, according to Fig. 8.3(c), the slope weighting with $P = 1/2$ has $Q_{max} = 3$ and $Q_{min} = 1/3$, where p_1 to p_5 are the allowable paths that satisfy the slope constraint, then, p_1 has the minimum slope of $1/3$ while p_5 has the maximum slope of 3.

The allowable region of Q_{max} and Q_{min} are defined as follows:

$$1 + \frac{\phi_x(k) - 1}{Q_{max}} \leq \phi_y(k) \leq 1 + Q_{max}(\phi_x(k) - 1) \tag{8.7}$$

$$M + Q_{max}(\phi_x - N) \leq \phi_y(k) \leq M + \frac{\phi_x(k) - N}{Q_{max}} \tag{8.8}$$

Eqn. 8.7 specifies the range of the points in the (i_x, i_y) plane that can be reached from the beginning point $(1,1)$. Similarly, Eqn. 8.8 specifies the range of points that have legal paths to the ending point (N, M). The overlapped region of Eqns. 8.7 and 8.8 form the allowable region. It should be noted that the allowable region might not exist if the regions described by Eqns. 8.7 and 8.8 are totally disjointed, i.e. $(M-1)/(N-1) > Q_{max}$ or $(M-1)/(N-1) < Q_{min}$, because it is unrealistic to compare two utterances that have very large time variations.

To solve Eqn. 8.2, the searching method DTW was used. DTW uses the dynamic programming technique [187] to search the optimal warping path ϕ for the dissimilarity function $d(\mathbf{X}, \mathbf{Y})$. This technique had been evaluated by many research groups [225, 232, 112] such that a recognition accuracy factor as high as 99.8% can be achieved [188]. However, due to the recursive approach of the dynamic programming technique, some restrictions on DTW, such as the stringent rule on slope weighting and the non-trivial finding of the K-best paths do exist. These restrictions have been raised by L.R. Rabiner and B.H. Juang in [173].

When the dynamic programming technique is used for solving the DTW problem, we encounter some practical problems. These problems can be regarded as follows:

1. *Exact endpoint time registration for utterance comparison* For most applications, the determination of the initial and final frames of an utterance is a highly imprecise calculation. Therefore, the endpoint constraints should be relaxed. We can use the following boundary constraints:

$$1 \leq \phi_x(1) \leq 1 + \Delta x, \quad N - \Delta x \leq \phi_x(T) \leq N \tag{8.9}$$
$$1 \leq \phi_y(1) \leq 1 + \Delta y, \quad M - \Delta y \leq \phi_y(T) \leq M \tag{8.10}$$

where Δx and Δy are the maximum anticipated mismatch or uncertainty in the endpoints of the patterns.

2. *Optimal time warping path as the exact solution for the time alignment in normalizing the temporal variability of utterances* The decision processing based on a single matching path is often too sensitive. It sometimes fails to cope with situations for minor deviations from the normal conditions. In some applications, there is more than one single matching path to be considered when a reliable decision is made.

3. *The assumption of the normalizing factor M_ϕ is independent of actual warping path* The normalizing factor for a weighted sequence is customarily the sum of the components of the weighting sequence, and takes the form

$$M_\phi = \sum_{k=1}^{T} m(k) \qquad (8.11)$$

In many cases, the above normalizing factor is a function of the actual path. This can be two of the four types as proposed by Sakoe and Chiba [188]. While it is possible to compute the normalizing factor for a given path, it makes the problem unwieldy if the minimization of Eqn. 8.2 is solved by the recursive dynamic programming algorithm.

Hence, it is not difficult to see that the DTW technique can suffer from the following drawbacks:

1. Stringent requirement of the slope weighting function;
2. Non-trivial finding n-best paths; and
3. Relaxed endpoint constraint.

8.4 Genetic Time Warping Algorithm (GTW)

Having established the essential arguments for the formulations of the time warping problem, the actual implementation procedure can then proceed. Considering the intrinsic properties of GA, a number of approaches may be adopted. The following subsections provide a detailed account of the methods used and the comparisons that have been made to iron out their differences.

To apply GA to solve the time warping problem, the optimal warping path must be mapped to the GA domain. These include the following considerations:

- a mechanism to encode the warping path ϕ as a chromosome.
- a fitness function to evaluate the performance of ϕ.
- a selection mechanism.
- genetic operators (crossover and mutation).

8.4.1 Encoding mechanism

In order to proceed with GTW, the warping path must be encoded in the form of a chromosome. Consider that a warping path $\phi' = (\phi'_x, \phi'_y)$ is represented by a sequence of points in the (i_x, i_y) plane, i.e.,

$$\phi' = (\phi'_x(1), \phi'_y(1))\ (\phi'_x(2), \phi'_y(2)) \cdots (\phi'_x(T), \phi'_y(T)) \qquad (8.12)$$

Eqn. 8.12 must satisfy the allowable and local constraints described in Sect. 8.3. The order of points must move along the allowable paths in which they are restricted by the slope constraint. Thus, it is preferable to express the warping function ϕ by a sequence of sub-paths. For example, the sub-path's $p_1, p_2, \ldots, p_5$ in Fig. 8.3(c) are encoded as $1, 2, \ldots, 5$. Therefore, the warping function ϕ can be expressed as follows:

$$\phi = (p_1)(p_2)\ldots(p_L) \qquad (8.13)$$

with initial points (i_{x_0}, i_{y_0}) and p_n being encoded as the allowable sub-paths.

There are several parameters that must be initialized, such as the global and local constraints of the warping function, in the initialization process. For example, Δx and Δy in Eqns 8.4 define the degree of relaxation in the global endpoints. P defines the degree of slope constraint. Q_{max} and Q_{min} in Eqns. 8.7 and 8.8 define the dimensions of the allowable region. Once the encoding scheme and the initialization of the system parameters have been completed, the population pool can be generated. The initialization procedures are summarized as follows:

1. Randomly select a beginning point;
2. Randomly select a sub-path;
3. Calculate the position of the ending point of the selected sub-path;
4. If the ending point of the selected sub-path falls outside the allowable region, then go to step 2;
5. Encode the gene as the selected sub-path. and absolute position of the ending point of the selected sub-path;
6. If the global ending points (n, m) has not reached, then go to step 2;
7. Repeat all steps until entire population is initialized.

8.4.2 Fitness function

Eqn. 8.1 is a distortion measurement between the two utterances X and Y. This provides the mechanism to evaluate the effectiveness of the warping function. However, the range of the distortion values calculated by Eqn. 8.1 sometimes can be unbounded. A fitness function has been designed to normalize Eqn. 8.1 to a range from 0 to 1. The normalized value of Eqn. 8.1 is the fitness of the warping path function which is used by the selection mechanism to evaluate the "survival-of-the-fitness" value of the warping function in subsequent generations. The procedures required to calculate the fitness values are as follows.

1. Calculate distortion values d_n of each warping function in the population by Eqn. 8.1, i.e. $d_n = d_{\phi_n}(\mathbf{X}, \mathbf{Y})$;
2. Find the maximum distortion value d_{max} in the population, i.e. $d_{max} = \max(d_1, d_2, \ldots, d_S)$;
3. Calculate the absolute differences dif_n between d_{max} and each d_n, i.e. $dif_n = d_{max} - d_n$;
4. Calculate the summation T of all the differences calculated in step 3, i.e. $T = \sum dif_n$; and
5. Calculate the fitness values f_n of each warping function by the equation $f_n = dif_n / T$.

8.4.3 Selection

The selection procedure is modelled after nature's own "survival-of-the-fittest" mechanism. Fitter solutions survive and weaker ones die. After selection, the fitter solutions produce more offspring and thus have a higher chance of surviving in subsequent generations.

GTW uses the Roulette Wheel selection scheme as its selection mechanism. In this type of selection, each solution is allocated a sector of a roulette wheel with the angle subtended by sector at the center of the wheel, which is equal to 2π multiplies by the fitness value of the solution. A solution is selected as an offspring if a randomly generated number in the range 0 to 2π falls into the sector corresponding to the solution. The algorithm selects solutions in this manner until the entire population of the next generation has been reproduced. The procedures to implement the Roulette Wheel selection scheme are outlined as follows:

1. Create an array $Sector$ with $S-1$ real numbers where S is the population size, i.e. real $Sector[1..S-1]$;
2. Set 1-st item of $Sector = 3600 \times$ fitness value of 1st warping path of the population, i.e. $Sector[1] = 3600 \times f_1$;
3. Iterate n from 2 to $S - 1$, set n-th item of $Sector$ = (n-1)-th item of $Sector + 3600 \times$ fitness value of n-th warping path in the population, i.e. in each iteration $Section[n] = Section[n-1] + 3600 \times f_n$;
4. Randomly select a number p from 0 to 3600;
5. Find index i of $Sector$ such that $Sector[i]$ is minimum in $Sector$ and $Sector[i] \geq p$. If i does not exist, set $i = S$. Then select the i-th warping path of the current population to the population of the next generation;
6. Repeat from step 4 until the entire population of next generation has been selected.

8.4.4 Crossover

The crossover procedure implements the exchange mechanism between two parent chromosomes. Crossover between selected fitter chromosomes in the population possibly reproduces a more optimized solution. The crossover rate is a probability factor used to control the balance among the rate of exploration, the new recombined building block and the rate of disruption of good individuals.

Since each chromosome of GTW represents a continuous warping path, arbitrary exchange between two chromosomes may generate two discontinuous warping paths. GTW has to ensure that the newly generated offspring remain in a continuous form. The procedures for crossover between two chromosomes are outlined as follows:

1. Randomly choose two warping paths A and B from the population;
2. Randomly generate a number p_c from 0 to 1. If $p_c >$ crossover rate then A will be selected as the offspring and finish;
3. Randomly choose a gene g_s from A, use the ending point stored in g_s as cross point s;
4. Search a gene g_e from B which has a ending point e such that point s can move to point e along an allowable path p_c. If no such gene exists then use A as the offspring and finish;
5. The offspring will be composed of two parts: *1*-st part is the segment of A from the *1*-st gene to g_s, *2*-nd part is the segment of B from the g_e to the last gene. Modify the allowable path in g_e of the offspring to p_c.

8.4.5 Mutation

The process of mutation randomly alters the genes of the chromosomes and takes on the role of restoring lost genetic material that has not been generated in the population during the initialization procedure. Because new genetic information will not be generated by the crossover procedure, the mutation procedure becomes an important mechanism by which to explore new genetic information.

Mutation is used to provide the random search capability for genetic algorithms. This action is necessary for solving practical problems that arise with multimodal functions. The mutation rate is also a probability factor which controls the balance between random searching and the rate of disruption of good individuals. It should be noted that the random alternation of genes during mutation may result in a discontinuous warping path. Therefore a special treatment of mutation is used in GTW to avoid this situation. The mutation procedures are summarized as follows:

1. Randomly choose a warping path A from the population;

2. Randomly generate a number p_m from 0 to 1. If $p_m >$ mutation rate then use A as the offspring and stop;
3. Randomly choose two genes g_s and g_e from A where g_s is positioned at a position ahead of g_e. The ending points stored in g_s and g_e marked as s and e, respectively.
4. Initialize a warping path p_m between points s and e;
5. The offspring will be generated by replacing the genes of A between g_s and g_e with p_m.

When all the GA operations are defined for the GTW problem, the evolutionary cycle of GA can then commence.

Now, it can be seen that the GA based DTW can solve the problems mentioned in the above section of the DTW. A number of points can be clarified here in order to bring about the best use of GTW:

1. the population size or the number of evaluations is constant in the GTW. This implies that the computational requirement is independent of the degree of endpoint relaxations;
2. the DTW solved by the dynamic programming technique is unable to compute the M_ϕ dynamically due to its recursive operation. This usually restricts the M_ϕ in Eqn. 8.1 to a constant;
3. GTW considers the solution of the warping function on a whole-path basis rather than a sub-path by sub-path basis. The M_ϕ used to evaluate the warping function can be obtained directly in each fitness evaluation;
4. the constraints of the slope weighting function $m(k)$ are relaxed and can be arbitrarily chosen in this case, such that M_ϕ need not to be a constant for all possible paths; and
5. the speech pattern distortions calculated in DTW are the minimum values of a number of traversed paths, and the Backtrack technique [Rabiner93] can only be used to compute a single optimal path so that it is difficult to obtain the second-best and the third-best warping paths.

The GA operates on a pool of populations and all the warping paths are stored as chromosomes in the pool. Therefore, warping paths can all be evaluated independently and K-best paths can be obtained naturally and without needing extra computational requirements.

8.4.6 Genetic Time Warping with Relaxed Slope Weighting Function (GTW-RSW)

We have pointed out that a comparison between two utterances that have large time differences is unrealistic. This problem can be alleviated by introducing a normalized slope weighting function on $m(k)$. This is possible when M_ϕ is computed by Eqn. 8.11.

However, the computation of M_ϕ can be very clumsy for DTW particularly when $m(k)$ is varied dynamically. Whereas for the case in GTW, in which each path is coded as a chromosome, then the computation of M_ϕ presents no problem to the GTW formulation. With such an unique property, the definition of $m(k)$ is therefore relaxed and can be chosen arbitrarily. In this way, a GTW scheme with a relaxed slope and weight function (GTW-RSW) can thus be performed.

8.4.7 Hybrid Genetic Algorithm

The GTW described above will produce results, and it is well known that GA does have the inclination to search optimally over a wide range of dynamic domains. However, it also suffers from being too slow in convergence. To enhance this searching capability and improve the rate of convergence, problem-specific information is desirable for GA so that a hybrid-GA structure is formed. In the present hybrid-GA formulation, we add problem-specific knowledge to the crossover operator so that reproduction of offspring that possess higher fitness values is realized.

In the hybrid-GTW, the hybrid-crossover genetic operator is proposed. The hybrid-crossover operator is similar to the original crossover operator whose procedure is listed as follows:

1. Randomly select two chromosomes A and B and perform the normal crossover operation. An offspring C is reproduced;
2. Swap chromosomes A and B and performs the crossover procedures again. Another offspring D is reproduced;
3. Instead of putting the offspring back to the population, a discrimination process is executed such that the best chromosomes among A, B, C, D will be put back to the population pool.

The experimental results of the hybrid approach of GTW are shown in next section indicating that the hybridized GTW can achieve better results than the tradition GTW using the same number of generations.

8.4.8 Performance Evaluation

To evaluate the performance of the above-mentioned schemes for the time warping problem, experimental results of the following four algorithms have been obtained:

1. Dynamic Time Warping Algorithm (DTW)
   The warping paths obtained by DTW are based on the dynamic programming searching method proposed by Rabiner [173];

2. Genetic Time Warping Algorithm (GTW)
   GTW used the traditional Genetic Time Warping technique described in Sect. 8.4. GTW used the $m(k)$ as defined in Eqn. 8.15, i.e. the normalization factor M_ϕ in Eqn. 8.1 must be constant for all warping paths;
3. Genetic Time Warping with Relaxed Slope Weighting function (GTW-RSW)
   GTW-RSW is the same as GTW except that the slope weighting function $m(k)$ used by GTW-RSW is relaxed. This means that the values of $m(k)$ can be arbitrarily chosen or $m(k)$ relaxed. M_ϕ can be varied for different warping paths; and
4. Hybrid Genetic Time Warping (Hybrid-GTW)
   Hybrid-GTW is the same as GTW-RSW except that it uses the hybrid-crossover operator described in Sect.8.4.7 instead of the traditional crossover operator.

A database of 10 Chinese words spoken by two different speakers was used with 100 utterances for each word. Each utterance was sampled at an 8.0KHz rate, 8-bits digitized and divided into frames of 160 samples. Ten-order cepstral analysis was applied as the feature measurement for the feature extractions. The initial and final endpoints for each word were determined by a zero cross rate and energy threshold. The short-time spectral distortion measurement is:

$$d(\mathbf{a}_R, \mathbf{a}_T) = \sum_{i=1}^{10} |a_{R_i} - a_{T_i}| \tag{8.14}$$

where $\mathbf{a}_R$ and $\mathbf{a}_T$ are the short-time spectral feature vectors of reference and test patterns, respectively.

For each of the 10 words, 80 arbitrarily chosen utterances act as the reference patterns while the remaining 20 utterances are used as test patterns. Each warping path in our experiments has five relaxed beginning points and ending points, i.e. the Δx and Δy in Eqns. 8.9 and 8.10 are set to five. The slope constraint with for P is set to $\frac{1}{2}$. The allowable region is defined as $Q_{max} = 3$ and $Q_{min} = 1/3$ in Eqns. 8.7 and 8.8. The following slope weighting function $m(k)$ for DTW is used:

$$m(k) = \phi_x(k) - \phi_x(k-1) + \phi_y(k) - \phi_y(k-1) \tag{8.15}$$

Table 8.1, summarizes the $m(k)$s used for the allowable step p_n for the DTW, while Table 8.2 shows the p_n for GTW, GTW-RSW, hybrid-GTW and parallel-GTW.

The population size for GTW and its derivatives is set to 40 chromosomes. The crossover rate is 0.6 and the mutation rate is 0.03333. The evolutionary cycle will be terminated at the end of the 40-th generations.

Table 8.1. Slope weighting function used in DTW

Allowable path used	$m(k)$ in Eqn. 8.1
p_1	4
p_2	3
p_3	2
p_4	3
p_5	4

Table 8.2. Slope weighting function used in GTW, GTW-RSW and Hybrid-GTW

Allowable path used	$m(k)$ in Eqn. 8.1
p_1	5
p_2	3
p_3	1
p_4	3
p_5	5

The experimental results of the four experiments are given as tabulated in Tables 8.3–8.6. On the basis of (Eqn. 8.1), the used symbols M_s, δ_s, M_d and δ_s are defined as the mean distortions for the same word, the standard deviations of distortions for same word, the mean distortions for different words and the standard deviations for different words, respectively.

Table 8.3. Experimental results of DTW

word	M_s	δ_s	M_d	δ_d
1	0.757	1.050	4.614	28.734
2	0.715	0.998	5.287	40.170
3	0.832	1.167	5.195	37.687
4	0.610	0.874	7.239	63.138
5	0.800	1.123	4.562	24.323
6	0.802	1.115	4.352	20.917
7	0.785	1.105	6.106	45.917
8	0.915	1.289	4.364	24.275
9	0.726	1.012	3.924	16.714
10	0.792	1.102	4.104	19.41

To assess the performance of the ASR system, a single value of the mean distortion M_s is not enough for classification. This is also compounds the fact that a variation of the slope weighting functions $m(k)$ (in Eqn. 8.1) has been used in the experiments.

Furthermore, the recognition rates for both DTW and GA approaches were found to be very close. Therefore, a more accurate measurement for the assessment uses the absolute difference between the M_s and M_d, i.e. $|M_s - M_d|$. This provides the discriminating abilities for recognizing confused

8.4 Genetic Time Warping Algorithm (GTW)

Table 8.4. Experimental results of GTW

word	M_s	δ_s	M_d	δ_d
1	1.125	1.670	5.475	38.98
2	0.959	1.362	6.136	52.335
3	1.322	2.142	5.985	47.339
4	0.789	1.101	8.202	79.839
5	1.202	1.861	5.443	33.958
6	1.244	1.944	5.210	29.477
7	1.092	1.638	7.024	59.473
8	1.328	2.143	5.107	31.867
9	1.088	1.603	4.629	22.954
10	1.321	2.133	4.954	27.30

Table 8.5. Experimental results of GTW-RSW

word	M_s	δ_s	M_d	δ_d
1	1.017	1.654	5.537	36.087
2	0.939	1.326	6.263	49.457
3	1.257	2.134	6.310	37.238
4	0.773	1.078	8.319	71.919
5	1.002	1.859	5.305	32.118
6	1.182	1.950	5.496	28.320
7	1.085	1.618	7.517	50.716
8	1.307	2.090	5.670	28.980
9	0.982	1.622	5.389	21.080
10	1.106	2.156	4.963	24.982

Table 8.6. Experimental results of Hybrid-GTW

word	M_s	δ_s	M_d	δ_d
1	0.911	1.638	5.599	33.191
2	0.909	1.290	6.390	46.579
3	1.192	2.126	6.635	27.137
4	0.757	1.055	8.436	63.999
5	0.802	1.857	5.167	30.278
6	1.120	1.956	5.782	27.163
7	1.078	1.598	8.010	41.959
8	1.286	2.037	6.233	26.093
9	0.876	1.641	6.149	19.206
10	0.891	2.179	4.972	22.657

216   8. Genetic Algorithms in Speech Recognition Systems

utterance, particularly for utterances with similar acoustic properties.

In this case, a lower value in $|M_s - M_d|$ implies that the ASR system has a weak ability to identify confused utterances, while a higher $|M_s - M_d|$ value implies that the ASR system has a high level of confidence in recognizing confused utterances. The results of $|M_s - M_d|$ the four experiments are tabulated in Fig. 8.4.

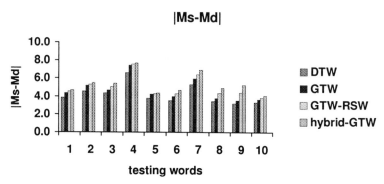

Fig. 8.4. The chart of $|M_s - M_d|$ of the four experiments

It can be clearly shown in the above figure that all the algorithms using GA technique have higher values of $|M_s - M_d|$ than those of DTW, and therefore a higher discrimination ability than the conventional DTW.

It is also expected that the hybrid approaches of the GTW should provide a faster convergence time. Figure 8.5 shows the results of Ms of GTW-RSW and hybrid-GTW. As shown in the figure, the hybrid-GTW has a smaller value of M_s than the GTW-RSW with the same number of generations in the evolutionary cycle. This implies that the hybrid-GTW has a faster convergence time than the GTW-RSW and further verifies that the use of hybrid-GTW can speed up the searching process.

8.5 Hidden Markov Model using Genetic Algorithms

Having demonstrated the effectiveness of applying genetic algorithms to the DTW problems, our next step is to focus on the HMM method. This approach can handle a large database much more efficiently than the DTW and has become an important statistical modeling method for automatic speech recognition (ASR) systems.

8.5 Hidden Markov Model using Genetic Algorithms

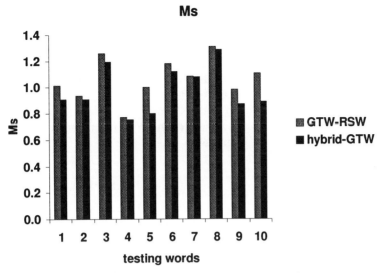

Fig. 8.5. The chart of Ms of GTW-RSW and Hybrid-GTW

HMM is able to determine the statistics of the variation of utterances from occurrence to occurrence. This capability makes HMM based speech recognizers more robust in comparison with the template matching methods, especially for the speaker-independent speech recognition.

The employment of HMM for ASR was proposed independently by Baker [11], the research group of IBM [8, 116, 117], and the HMM related researches are still active. One of the major objectives in the HMM research is to find a good HMM for the best described spoken word. There are two major issues to be solved before reaching such an objective. The first is to determine the topology of the HMM, and the second is to optimize the model parameters such that they can represent the training utterances accurately.

Thus far, we have yet to reach a simple but theoretically correct way of making a good choice of topology for HMMs, and this usually is application specific. In particular, the class of topology for HMM called left-to-right model [13], this is generally used for isolated word recognition.

There are several ideas involved in determining the number of states of HMM for each word. Levinson [130] suggested that the number of states should roughly correspond to the number of sound units in the word. On the other hand, Bakis [13] suggested that the number of states should match the average number of observations in the spoken version of the word as more appropriate. Due to the variability of utterances, the determination of the optimal number of states used in the word model still remains unsolved, and

218   8. Genetic Algorithms in Speech Recognition Systems

there is still no golden rule for the magic number of states for HMM.

Many successful heuristic algorithms such as the Baum-Welch algorithm [15, 16, 17] and the gradient methods [117] for the optimization of the model parameters do exist. However, these methods are all hill-climbing based algorithms and dependent quite strongly on the initial estimates of the model parameters. In practice, this approach usually leads to a sub-optimal solution. While this problem can be avoided by the segmental K-means segmentation method [173], it is computationally intensive.

8.5.1 Hidden Markov Model

To be able to apply the HMM as the commonly used speech recognition system, there are three major components to be considered:

1. a number of finite states used to describe an utterance, i.e. an ergodic or fully connected HMM to allow any state of the HMM to be reached by any other states; or this could be simply a left-to-right HMM;
2. the transition probability matrix that represents the probability function of the transition among states in the HMM and
3. the observation symbols probability matrix that represents the probability for all possible symbols in each transition state.

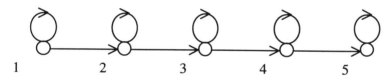

Fig. 8.6. A five states left-to-right HMM

Fig. 8.6 shows a simple example of a five-states left-to-right HMM. The HMM states and transitions are represented by node and arrows, respectively. There are five states in the HMM model. Thus, a state matrix of (5 $times$ 5) elements can be identified. Since this model is not an ergodic model, only 9 transition cases (arrows) are existed. Therefore, a 5 $times$ 5 states matrix with 9 non-zero elements is identified. This state matrix is usually called the **A** matrix. In each time interval for any transitions exist, an observation symbol will be emitted. Thus, a sequence of observation symbols will appear over a finite time interval. These symbols are used for the representation of a speech signal. In general terms, these are the symbols for the acoustic feature vectors of speech utterances, such as the ceptral or linear predictive coefficients.

8.5 Hidden Markov Model using Genetic Algorithms

For each observation symbol, there is a probability associated with each state transition. A probability density function is usually called the **B** matrix in HMM terminology. For example, in a transition that has 256 possible observations, then the B matrix will equal to 9×256. When the HMM is fully connected, the matrix is therefore 25×256 in size. Each entry will have a probability associated with it. The **B** matrix, can be in the form of a discrete distribution or a continuous distribution and is also known as the discrete HMM and continuous HMM, respectively.

Thus, in the training mode, the task of the HMM-based speech recognition system is to find a HMM model that can best describe an utterance (observation sequences) with parameters that are associated with the state model **A** and **B** matrix. A measure in the form of maximum likelihood is computed among the test utterance and the HMMs, and the one with the highest probability value against the test utterance is considered as the candidate of the matched pattern.

The problem can best be tackled by the use of GA in combination with the HMM. With the proper design of chromosomes, the GA-HMM approach can be modified to solve problems of a similar nature with even greater complicity for a smaller effort.

8.5.2 Training Discrete HMMs using Genetic Algorithms

The advantage of using GA for HMM training is its ability for global searching. It provides a good mechanism to escape from the local maxima or at least provides a better local maxima. To illustrate this principle for HMM [28], a full account of GA-HMM model formulation is given below.

Discrete Hidden Markov Model. A discrete HMM λ is characterized by the following parameters and these are formally defined as follows:

1. N — The number of states in the model. This parameter is pre-defined at the beginning of the problem and is hard coded in the implementation. For example, the HMM model for a phonetic recognition system will have at least three states which correspond to the initial state, the steady state, and the ending state, respectively;
2. M — The number of distinct observation symbols per state are considered as the feature vector sequence of the model. This number can be large as the speech utterances are continuous in nature. As for a discrete HMM, a limited number of distinct symbols is reached via vector quantization although the parameters may be pre-defined;
3. $\mathbf{A} = \{a_{ij}\}$ — the state transition probability distribution, where

$$a_{ij} = P[q_{t+1} = j | q_t = i] \quad 1 \leq i, j \leq N \quad (8.16)$$

is the probability of state transition from i to j;

4. $\mathbf{B} = \{b_j(k)\}$ — the observation symbol output probability distribution, where

$$b_j(k) = P[o_t = v_k | q_t = j] \quad 1 \leq j \leq M \quad (8.17)$$

is the output probability of symbol v_k emitted by state j; and

5. $\pi = \{\pi_i\}$ — the initial state distribution, where

$$\pi_i = P[q_1 = i] \quad (8.18)$$

Having defined the necessary five parameters that are required for the HMM model, the actual computation of the probability distribution A, B, and π is only required., i.e.

$$\lambda = (\mathbf{A}, \mathbf{B}, \pi) \quad (8.19)$$

It should be noted that the number of states and the number of distinct symbols are hidden but implicitly used within the model.

Encoding Chromosome Mechanism. Considering that the type of the HMM is pre-fixed in advance, then, a five states left-right HMM model with a set of 256 observation symbols in each state is shown in Fig. 8.6.

The parameters of the HMM model consisted of two matrices: $\mathbf{A}$ and $\mathbf{B}$ where Matrix $\mathbf{A}$ is a 5 $times$5 transition probability distribution matrix, whose element at row i, column j are the probability $a_{i,j}$ of transition from the current state i to the next state j must satisfy the following condition:

$$1 = \sum_{j=1}^{5} a_{i,j} \quad \text{where} \quad i = 1, \ldots, 5 \quad (8.20)$$

Matrix $\mathbf{B}$ is a 5×256 observation symbol probability distribution matrix, whose element at row i and column k is the probability $b_{i,k}$ of observation symbol with index k emitted by the current state i which also must satisfy the condition:

$$1 = \sum_{k=1}^{256} b_{i,k} \quad \text{where} \quad i = 1, \ldots, 5 \quad (8.21)$$

In the GA-HMM training, the model is encoded into a string of real numbers, as shown in Fig. 8.3. This string acts as a chromosome composed of two parts: $\mathbf{A}$' and $\mathbf{B}$'. These are formed by concatenating the rows of matrices $\mathbf{A}$ and $\mathbf{B}$ respectively. Due to the configuration of the model, some transitions between the states do not exist. Therefore, the corresponding elements in matrix $\mathbf{A}$ always have a zero value and are not encoded as genes

8.5 Hidden Markov Model using Genetic Algorithms

of the chromosomes. As a result, a total of 9 elements (genes) of matrix **A** are coded as a chromosome.

The size of the **B** matrix then consists of 5×256 none-zero elements in the matrix. Thus, a HEM chromosome is formulated as a combination of the **A** and **B** matrices as shown in Fig. 8.7.

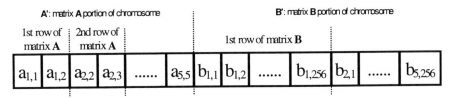

Fig. 8.7. Genetic chromosome in a GA-HMM model

Selection Mechanism. GA-HEM training uses the Roulette Wheel selection scheme as its selection mechanism. In the selection, each chromosome is allocated a sector of the Roulette Wheel which is equal to 2π multiplied by the fitness value of the solution. An offspring is selected if a randomly generated number in the range 0 to 2π falls into the sector corresponding to the solution. This selecting scheme continues in this manner until the entire population of the next generation has been reproduced. This mechanism is similar to the one used for GTW.

Crossover and Mutation. In the GA-HEM training, both one point and two-point crossover are used. This particular combined arrangement enables multiple point crossover to facilitate the entire matrices of **A** and **B**. In this way, the chromosome information will be retained and interchange between the two matrices will be restricted.

A typical example of of the crossover operation is shown in Fig. 8.8. The two parent chromosomes are randomly selected from the population pool using random real number between 0 and 1 for each gene. This number is then used to compare against the crossover rate. If the number is greater than the crossover rate, the two parents act as offspring and no crossover occurs. Otherwise, one cross point and two cross points will be randomly applied to the portions of **A'** and **B'** respectively. The portions of the chromosomes between the selected cross points are exchanged to generate new chromosomes.

The mutation operation in GA-HMM training alters three real numbers at a time. Mutation selects one chromosome randomly from the population

222   8. Genetic Algorithms in Speech Recognition Systems

pool and designates it as the parent. Similar to the crossover operator, it uses the comparison between the randomly generated number and mutation rate for the determination of the mutation operation. When the decision is positive, the three probabilities will select the parent randomly. Again, one is for the part **A'** and the other two are for the part **B'**. These three selected probabilities will replace the three random generated real numbers.

When the crossover or mutation operations have been completed, the chromosome must be normalized in order to satisfy the constraints mentioned in Eqns. 8.20 and 8.21.

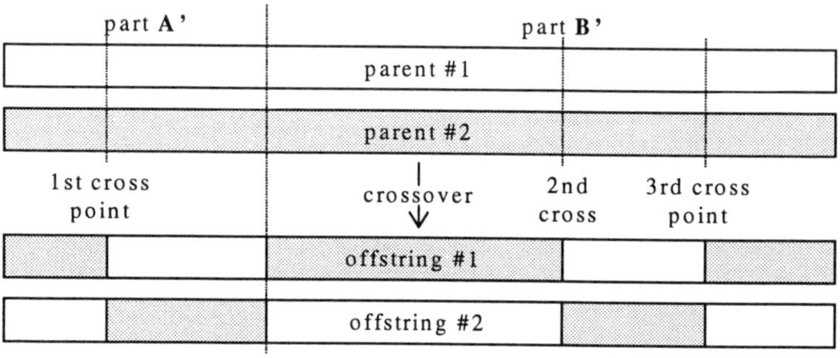

Fig. 8.8. Three-point crossover operation

Fitness Value Evaluation. The average probability p_n of the HMM solution λ_n that generates the training observation sequences $O_1, \ldots, O_M$ are used as references for the fitness evaluation:

$$p_n = \frac{1}{M} \left(\sum_{i=1}^{M} p(O_i|\lambda_n) \right) \quad (8.22)$$

where $p(O_i|\lambda_n)$ is calculated by a forward-backward procedure [173]. The fitness value f_n of solution λ_n is calculated as follows:

$$f_n = \frac{p_n}{\sum_{i=1}^{N} p_i} \quad (8.23)$$

where N is number of solutions in the population.

Experimental Results. To demonstrate the proposed scheme, the results obtained by the GA-HMM training method are used for comparison with those trained by a forward-backward procedure [173]. The initial model parameters are randomly generated and the constraints in Eqns. 8.20 and 8.21 are enforced. Ten experiments were carried out. In each experiment, two

8.5 Hidden Markov Model using Genetic Algorithms

HMMs were trained with the same ten observation sequences for both the GA-HMM and the forward-backward procedures. In the GA-HMM training, the following control parameters were used:

Table 8.7. Control parameter for GA-HMM

Population size	30
Crossover rate	0.01
Mutation rate	0.0001

The required two values of the HMM are computed:

1. p_{same} — average log probability of the HMM generated by the 10 training observation sequences of the same HMMs; and
2. $p_{different}$ — average log probability of the HMM generated by the 90 training observation sequences of the different HMMs.

Throughout the experiment, the HMM training using the forward-backward procedure was terminated after 200 turns or where the increase of average log probability p_{same} was less than 0.00001. As for the GA-HMM training, the termination would take place when 20000 evolution cycles was reached. The obtained experimental results are listed in Table 8.8.

Table 8.8. Experimental results: p_{same} and $p_{different}$

Experiment	Genetic Algorithm		Forward-Backward Procedure	
	p_{same}	$p_{different}$	p_{same}	$p_{different}$
#1	-4.9473	-7.4982	-4.7359	-7.2714
#2	-3.5693	-8.9727	-4.2125	-8.6137
#3	-3.2932	-8.6473	-4.9843	-7.5914
#4	-3.0982	-8.5291	-4.3908	-7.7634
#5	-4.2345	-9.1483	-4.3876	-7.1007
#6	-3.3281	-7.5581	-4.9811	-7.3825
#7	-4.1869	-7.6257	-4.3481	-7.7351
#8	-4.2322	-8.6274	-4.0567	-7.9328
#9	-4.3872	-8.7812	-4.4860	-7.7514
#10	-3.1539	-8.3641	-4.9251	-8.2254

As indicated in Table 8.8, the HMMs trained by the GA-HMM method has higher average log probabilities p_{same} than those obtained by the forward-backward procedure, except for the experiments of #1 and #8. This means that, in most cases, the obtained HMMs trained by GA-HMM training are superior to those trained by the forward-backward procedure most of the times.

224   8. Genetic Algorithms in Speech Recognition Systems

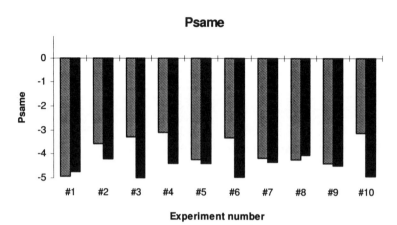

Fig. 8.9. The chart of p_{same} of GA-HMM and forward-backward procedure

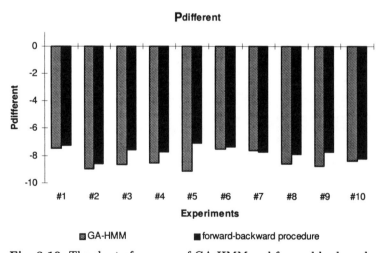

Fig. 8.10. The chart of $p_{different}$ of GA-HMM and forward-backward procedure

In the case of experiments #1 and #8, GA failed to reach the required solution even after 20,000 generations. This occurred in spite of the fact that the average log probabilities $p_{different}$ of the HMMs trained by GA has lower values. This indicates that the models are more optimized to the training observation sequences of the same word only, but not the training observation sequences for other words. This also implies that the HMMs trained by GA-HMM has a higher discrimination ability.

The above results indicated that the GA-HMM training method has a higher probability of finding a better solution or at least a better local maximum. It also shows that a better HMM model can be generated by GA-HMM as compared with that obtained by the forward-backward algorithm.

8.5.3 Genetic Algorithm for Continuous HMM Training

In this domain, we further extend our work to continuous GA-HMM, in which the matrices of **A** and **B** are not only optimally obtained, but can also determine the exact number of states of the HMM model in an optimal fashion.

Because of this unique feature, the GA operation such as the encoding mechanism, the fitness evaluation, the selection mechanism, and the replacement mechanism have to be re-designed in order to accommodate the extra requirement. In addition, as with GTW, where the hybrid-GA operations have been employed to enhance the convergence speed, the same approach is also adopted here together with the application of the Baum-Welch algorithm [16].

Continuous distribution HMM. The formal specifications of a HMM with continuous observation densities are characterized by the following model parameters and we used the same notation as described by Rabiner [173]. This is very much the same as the discrete HMM except that the finite mixture of the Gaussian distribution function is used as the output observation symbol probability function.

1. N, the number of states in the model
2. M, the number of mixtures in the random function
3. **A**, the transition probability distribution matrix $\mathbf{A} = \{a_{ij}\}$, where a_{ij} is the transition probability of the Markov chain transiting to state j, given the current state i, that is

$$a_{ij} = P[q_{t+1} = j | q_t = i] \quad (8.24)$$

where q_t is the state at time t and a_{ij} has the following properties:

$$a_{ij} \geq \phi \quad 1 \leq i, j \leq N \tag{8.25}$$

$$\sum_{j=1}^{N} a_{ij} = 1 \quad 1 \leq i \leq N \tag{8.26}$$

4. For the continuous distribution HMM model, the observation symbol probability distribution matrix $\mathbf{B} = \{b_j(\mathbf{o})\}$ is defined below where $b_j(\mathbf{o})$ is the random function associated with state j. The most general representation of the random function is a finite mixture of Gaussian distributions which has a form as follows:

$$b_j(\mathbf{o}) = \sum_{k=1}^{M} c_{jk} G(\mathbf{o}, \mu_{jk}, \mathbf{U}_{jk}), \quad 1 \leq j \leq N \tag{8.27}$$

where $\mathbf{o}$ is the observation vector, c_{jk} is the mixture coefficient for the k-th mixture in state j, and G is the Gaussian distribution with mean vector μ_{ik} and covariance matrix $\mathbf{U}_{jk}$ for the k-th mixture component in state j.

The mixture coefficient c_{jk} has the properties that satisfy the following stochastic constraints:

$$c_{jk} \geq \phi \quad 1 \leq j \leq N, \ 1 \leq k \leq M \tag{8.28}$$

$$\sum_{k=1}^{M} c_{jk} = 1 \quad 1 \leq j \leq N \tag{8.29}$$

5. π, the initial state distribution matrix $\pi = \{\pi_i\}$ in which

$$\pi_i = P[q1 = i] \quad 1 \leq i \leq N \tag{8.30}$$

It can be seen that the elements of a HMM include the model parameters of N, M, $\mathbf{A}$, $\mathbf{B}$, and π. However, the values of N and M exist implicitly as indicated by the dimension of the matrices $\mathbf{A}$ and $\mathbf{B}$ respectively. Similar to the discrete HMM, the following notation can be used to represent the HMM.

$$\lambda = (\mathbf{A}, \mathbf{B}, \pi)$$

Encoding Mechanism. An appropriate encoding mechanism allows us to convert the required solution in the form of chromosomes. This representation becomes a very important data structure for the implementation of GA. However, when a bad data structure is used, particularly for crossover operations, the arrangement of the chromosomes will complicate the overall genetic operations which in the end, will cause difficulty in translating

8.5 Hidden Markov Model using Genetic Algorithms

the results. In this case, we adopt the decode of phenotype representation in our encoding mechanism. This takes the form of a hierarchical data structure which has a convenient data format specifically designed for fitness evaluation. This typical phenotype format is shown in Fig. 8.11.

```
structure {
    real N;        /* number of states in the model */
    real Mat_A[Nmax][Nmax];   /* transition probability distribution matrix: A,
                                 element Mat_A[i][j] corresponding to the
                                 transition probability a_ij */
    structure {
        real c;        /* mixture coefficient */
        /* Gaussian probability density function */
        real Mean_Vector[Vdim];              /* mean vector: μ */
        real Covariance_Matix[Vdim][Vdim];   /* covariance matrix: U */
    } Mat_B[Nmax][M];  /* observation symbol probability distribution matrix: B,
                         element Mat_B[j][k] corresponding to the k-th mixture
                         component of the random function of state j */
} HMM;    /* top level data structure of HMM */
```

Fig. 8.11. Data structure of phenotype used in the GA-HMM training

$Nmax$ is the maximum number of states allowed in each word model and the constant $Vdim$ is the size of the observation vector, and the other variables correspond to the elements of the HMM described in Sect. 8.5.3.

The chromosome in traditional GA is usually expressed in a string of elements and are termed as genes. According to the problem specifications, the genes can be in the type of binary, real number, or the other forms. Bit-string encoding is a classic approach which generally is used by GA researchers due to its simplicity and tractability. Since the basic data type of the elements of the HMM are real numbers, this format is adopted for the representation of chromosomes as shown in the following figure.

Where $mv_{i,j,k}$ is the k-th scalar in the mean vector of the j-th mixture of the random function in state i and $cv_{i,j,k,l}$ is the element in k-th row and l-th column of the covariance matrix of the j-th mixture of the random function in state i. It can also be found that the size of chromosomes is governed by the number of states in the HMM model. The larger the number of states N,

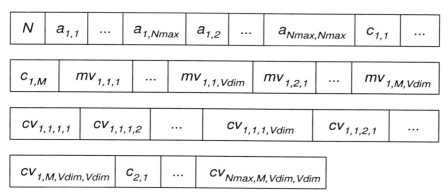

Fig. 8.12. The encoding of chromosomes used in the GA HMM training

the longer the chromosomes.

Fitness Evaluation. The fitness evaluation is a mechanism used to determine the degrees of the optimized solutions to the problem. Usually, there is a fitness value associated with each chromosome. A higher value of fitness value means that the chromosome or solution is more appropriate to the problem while a lower value of fitness value indicates a lesser count.

In our GA formulation, the fitness values are the objective functions. As previously described in discrete GA-HMM, the maximum likelihood $P[\mathbf{O}|\lambda]$ is a suitable criterion used for the objective function for the determination of the quality of the chromosomes. However, the dynamic range of the $P[\mathbf{O}|\lambda]$ is usually very small, i.e. $P[\mathbf{O}|\lambda] < 10^{-200}$, and often exceeds the precision range of any machines (even in double precision). So instead, we use the logarithm of the likelihood.

Then, the objective function is defined as the average of the logarithms of the probabilities of the observation sequences, $\mathbf{O}_1, \ldots, \mathbf{O}_S$ generated by the given n-th HMM λ_n or n-th chromosome in the population. This has the form:

$$lp_n = \frac{1}{S}\left(\sum_{i=1}^{S} \log\left(P[\mathbf{O}_i|\lambda_n]\right)\right) \quad (8.31)$$

where lp_n is the fitness value of the n-th chromosome in the population, the likelihood $P[\mathbf{O}|\lambda]$ is calculated by the forward procedure [Rabiner 93] and S is the number of observation sequences in the training data.

Selection Mechanism. The well known Roulette Wheel Selection is used as the selection mechanism. Each chromosome in the population is associated

8.5 Hidden Markov Model using Genetic Algorithms

with a sector in a virtual wheel. According to the fitness value of the chromosome, which is proportional to the area size of the sector, the chromosome that has a higher fitness value will occupy a larger sector whilst a lower value takes the slot of a smaller sector. Figure 8.13(a) shows a typical allocation of five sectors of chromosomes in the order of 3,2,6,1 and 4.

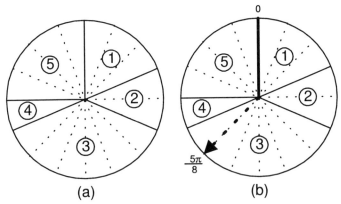

Fig. 8.13. (a) Example of sector allocation in the roulette wheel selection (b) Example of selection by the roulette wheel selection

Once the sector allocation is completed, a vertical line is placed on the wheel which can be arranged next to the sector where the highest value of chromosome is located. This vertical line is marked as a reference point attached with a zero radian. To select a parent, the straight-line maker rotates around the circumference of the wheel according to a randomly generated angle between 0 to 2π radian from the reference vertical line. This is usually done by a uniform random number generator. A chromosome is selected if the marker is placed on the associated sector of the chromosome. Figure 8.13(b) is an example for the random generated angle being equal to $5\pi/8$ radian. In this case, the chromosome in the 3-rd sector chromosome is selected. It can be seen that a chromosome in a larger sector, or of a higher fitness value, will have a higher chance of being selected. In practice, this selection process also has to comply with the constraints stated in Eqn. 8.33, where a fitness value normalization based on Eqn. 8.32 should apply.

$$Nlp_n = \frac{2\pi \left(lp_n - \min_{1 \leq j \leq P}(lp_i)\right)}{\sum_{j=1}^{P} \left(lp_j - \min_{1 \leq j \leq P}(lp_i)\right)} \qquad 1 \leq n \leq P \qquad (8.32)$$

$$Nlp_n \geq 0, \qquad 1 \leq n \leq P$$

$$\sum_{n=1}^{P} Nlp_n = 2\pi \tag{8.33}$$

where Nlp_n is the normalized fitness value of the n-th chromosome in the population and P is the number of chromosomes in the population. A random number r between 0 to 2π is uniformly generated uniformly and that the n-th chromosome is the parent if n is the minimum value of the argument k of the statement shown in Eqn. 8.34.

$$\sum_{i=1}^{k} Nlp_i \geq r, \quad 1 \leq k \leq P \tag{8.34}$$

Genetic Operations. Genetic operations are the essential and major tools for GA. Here, four genetic operators are designed to perform such operations. Figure 8.14 shows the execution sequence of the genetic operations.

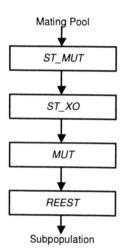

Fig. 8.14. Execution sequence of genetic operators in our GA-HMM training

These operations are termed as the state mutation (ST_MUT), state crossover (ST_XO), mutation (MUT), and re-estimation ($REEST$) operators.

State Mutation (ST_MUT). State mutation is the only tool in this formulation that can change the state number in the chromosomes. It aims to explore the fitness of the chromosomes with a different number of states.

In this operation, the number of states in the HMM is determined. A uniform random number generator is used to generate a random number in

8.5 Hidden Markov Model using Genetic Algorithms 231

the range of 0 to 1. If the random number is larger than 0.5, then the number of states generated in the offspring N_o equals to the number of states in the parent N_p plus 1 (i.e. $N_o = N_p + 1$). Otherwise, N_o equals N_p minus 1 (i.e. $N_o = N_p - 1$). This is followed the finding of the fittest chromosome within the number of states N_o in the population. Should such a chromosome not exist, a chromosome within the number of states N_o is randomly generated. The MUT will modify either the selected chromosome or the randomly generated chromosome in a random fashion, and the resultant chromosome is then identified as the offspring. This is a very simple genetic operator, but has the capability to efficiently find the best number of states in the word model.

State Crossover (ST_XO). State crossover operator is a derivative of the standard crossover operator. It combines sub-parts of the two parents to produce offspring that can preserve both parents' genetic material. In this operator, a parent is selected to match the fittest chromosome that has same number of states in the population pool. If this chromosome is not found, then MUT will be applied to an original chromosome(single parent) and this becomes identified as the second parent. Three states are randomly selected from the second parent. The offspring are reproduced by mutually interchanging the corresponding states in the first parent with the selected states. Figure 8.15 is an example of ST_XO where the states of two parents have been recombined with the crossover points at state 2, state 4, and state 5 respectively.

It should be noted that we used 'state' instead of 'model parameters' as the crossover unit. This is because the exchange of partial state will make the fitness value of the chromosome drop significantly (i.e. $lp_n < -4000$) and render the offspring unusable. This may due to the correlation between model parameters within a state being destroyed by changing the state information partially.

Mutation (MUT). Mutation introduces variations of the model parameters into chromosomes. It provides a global searching capability for our GA by randomly altering the values of genes in the chromosomes. It recovers the information lost in the initialization phase and enable the correct genes to escape from the initial conditions and gradually converges to the required model parameter setting.

According to an alternation probability, each model parameter in the parent may or may not be altered by this genetic operator. Before the modification of a model parameter, the alternation probability is compared with a randomly generated probability to test if the alternation probability is larger than or equal to the randomly generated probability. If this is positively

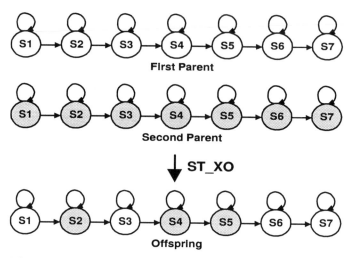

Fig. 8.15. Example of ST_XO with the crossover points at state 2, state 4, and state 5 respectively

factored in, then the model parameter x_{λ_n} is altered by the equation as follows:

$$x_{\lambda_n} = x_{\lambda_n} \times G(1.0, 0.001) \tag{8.35}$$

where $G(1.0, 0.001)$ is a Gaussian random number generator with mean=1.0 and variance=0.001. If the result is false, then the model parameter will not be changed.

Re-estimation (REEST). This is used to improve the fitness values of the offspring. The Baum-Welch algorithm with three iterations is applied to the offspring. It should be noted that the offspring from the *MUT* and *ST_MUT* are generated by random alteration. The model parameters of the offspring may then violate the stochastic constraints in Eqns. 8.25, 8.26, 8.28 and 8.29. Therefore, in order to satisfy the stochastic constraints before the application of the re-estimation formulas, the randomly altered chromosomes should first be normalized.

The normalization of the offspring will limit the lower bounds of the model parameters so that the Eqns. 8.25 and 8.28 can both be satisfied. If any model parameter is smaller than its associated limit, then the model parameter will be set to that limit. The associated limits for the transition probability a_{ij}, the mixture coefficient c_{jk}, and the elements of the covariance matrix $\mathbf{U}_{jk}$ are set to 0, 0.01, and 0.0001 respectively.

8.5 Hidden Markov Model using Genetic Algorithms

Then, the scheme to normalize the transition probability a_{ij} and the mixture coefficient c_{jk} to satisfy the Eqns. 8.26 and 8.29 may be observed by the following equations.

$$\begin{align}
\bar{a}_{ij} &= \frac{a_{ij}}{\sum_{n-1}^{N} u_{in}}, & 1 \leq i \leq N \\
\bar{c}_{jk} &= \frac{c_{jk}}{\sum_{m=1}^{M} c_{jm}}, & 1 \leq j \leq M
\end{align} \tag{8.36}$$

where $\bar{a}_{ij}$ and $\bar{c}_{jk}$ are the normalized values of a_{ij} and c_{jk} respectively, N is the number of states in the chromosome, and M is the number of mixings used in the random function.

Replacement Strategy. After the sub-population (pool of offspring) pool is generated, the offspring must be inserted back to the population pool. In our GA, the Steady-State Reproduction scheme is used as the replacement strategy. According to this strategy, the newly generated population for the next generation will replace the worst chromosomes in the current population.

Hybrid-GA. Given that with the large size of feature vector as is usually applied to the GA-HMM algorithm, a successful completion within a short terminating time is very difficult to achieve. The use of hybrid-GA improves the convergence capability of the population pool and can alleviate the computation burden of GA operations. The flow of this scheme of operations is shown in Fig. 8.16. The idea is such that for every ten generations produced, the hybrid-GA employs the Baum-Welch algorithm with eight iterations to improve the fitness value of each chromosome in the population pool. In this way, better parents will be selected for the next generation which, in turn, improves the overall GA cycles.

Experimental Set-up. To verify the proposed GA-HMM based on the Baum-Welch algorithm, the HTK Toolkit V1.5 [242] was adopted, and the re-estimation tool was also extracted from it. The Hrest for HMM training and the initialization tool (Hinit) for each HMM were utilized. The training process terminated when the increase of the lp between two successive iterations was less than 0.0001, and at the same time, the GA evolution finished after 30 generations.

The experimental results are judged in terms of the average of the logarithms probabilities lp defined in Eqn. 8.33. A total of one hundred words was extracted from the TIMIT Corpus for the data training. Each word was sampled at 16,000 Hz with 16-bit resolution. Twelfth order mel-frequency

234   8. Genetic Algorithms in Speech Recognition Systems

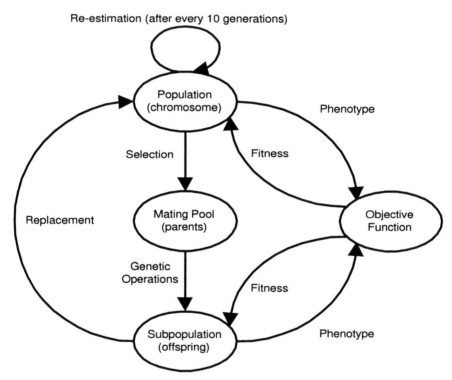

Fig. 8.16. Hybrid-GA for HMM training

8.5 Hidden Markov Model using Genetic Algorithms

cepstral coefficients were used as the feature vectors, while the left-to-right model with four mixture components in the random function was adopted for HMM training. The number of states in each word model was set in the range from 7 to 10. The control parameters used are shown in Table 8.9:

Table 8.9. Control parameters for GA-HMM

Population Size	30
Sub-population Size	3
State Mutation Rate	0.6
State Crossover Rate	0.6
Mutation Rate	0.05
Alternation Probability	0.01

Performance of our GA-HMM Training. The experimental results are listed in Tables 8.10 and 8.11. In the GA-HMM column, the values in the parenthesis indicated the best available number of states found by GA-HMM training and the italics value in each row is the best result obtained from the Baum-Welch algorithm. As can be seen from Tables 8.10 and 8.11, the proposed GA-HMM training is able to find the number of states with best results in all one hundred HMM trainings. In addition, the trained HMMs via GA-HMM have the higher values of lp than those HMMs derived by the Baum-Welch algorithm alone. This means that the GA-HMM method is the superior of the two.

Remarks. The attractiveness of this GA-HMM for speech recognition, is undoubtedly due to its power to obtain both HMM model parameters and the lowest possible of number of states in a speedy simultaneous manner during the training process. This achievement can be further demonstrated by the following selected examples. A total of four different words, namely word #3, word #12, word #35 and word #48 were randomly chosen for this purpose and for each word used, the chromosomes with a number of state, i.e. 7, 8, 9 and 10 were assigned. The outcome of the GA-HMM training was then judged by its ability to identify suitable chromosomes that fulfilled the required criteria.

Based on the results that are shown in Figs. 8.17 to 8.19, the distribution of the number of occurrences in the population pool for each type of the chromosomes were evenly spread, although this was done by a random allocation process. However, the pool tended to be dominated by a specific chromosome with an appropriate number of states, while the others started to disappear gradually from the population as the generation progressed. This effect is largely contributed by the combination of the state mutation operator and Roulette Wheel selection. This turned out to be a significant

Table 8.10. Experimental results (#1 — #50): lp_n

Word	GA-HMM	Re-estimation (7 states)	Re-estimation (8 states)	Re-estimation (9 states)	Re-estimation (10 states)
#1	-210.63(10)	-256.89	-243.42	-236.72	*-214.00*
#2	-307.49(10)	-317.05	-313.57	-310.87	*-307.58*
#3	-255.99(10)	-292.46	-287.84	-281.22	*-263.22*
#4	-295.02(10)	-310.10	-303.71	-301.40	*-295.05*
#5	-257.38(10)	-285.47	-271.44	-262.69	*-257.44*
#6	-180.24(10)	-216.24	-206.66	-196.27	*-180.37*
#7	-269.94(10)	-281.06	-278.36	-276.09	*-269.94*
#8	-277.85(10)	-293.74	-288.65	-281.46	*-277.85*
#9	-305.77(10)	-329.29	-319.15	-312.24	*-305.77*
#10	-312.00(10)	-322.55	-318.07	-314.55	*-312.02*
#11	-261.64(10)	-278.06	-271.57	-269.25	*-261.64*
#12	-227.97(9)	-255.92	-251.18	*-229.13*	-234.09
#13	-219.25(10)	-260.30	-237.93	-225.31	*-219.25*
#14	-325.60(10)	-353.65	-340.79	-337.23	*-326.48*
#15	-286.22(10)	-296.85	-293.87	-290.51	*-288.08*
#16	-197.11(10)	-229.24	-221.73	-210.99	*-203.92*
#17	-249.99(10)	-279.42	-259.52	-257.78	*-253.04*
#18	-273.39(10)	-286.52	-284.19	-280.86	*-275.35*
#19	-316.26(10)	-321.56	-321.86	-318.10	*-316.26*
#20	-306.64(10)	-316.84	-311.14	-308.31	*-306.66*
#21	-169.66(10)	-216.20	-215.60	-196.27	*-172.39*
#22	-260.50(9)	-279.48	-273.09	*-265.15*	-266.04
#23	-192.10(10)	-219.89	-209.93	-198.29	*-198.02*
#24	-295.74(10)	-316.95	-310.21	-301.75	*-296.82*
#25	-243.14(10)	-268.13	-257.91	-249.72	*-244.31*
#26	-242.05(10)	-279.79	-276.77	-269.58	*-246.76*
#27	-279.98(10)	-298.67	-289.89	-283.95	*-279.98*
#28	-275.05(10)	-283.45	-281.45	-278.62	*-275.17*
#29	-210.03(10)	-246.21	-228.50	-234.02	*-210.19*
#30	-104.56(10)	-143.08	-145.96	-147.41	*-108.89*
#31	-290.55(10)	-308.09	-301.27	-295.87	*-290.65*
#32	-113.97(10)	-170.52	-151.46	-128.26	*-116.48*
#33	-284.24(10)	-300.00	-293.23	-286.70	*-285.00*
#34	-179.70(9)	-210.78	-209.95	*-188.00*	-188.03
#35	-244.91(10)	-271.27	-265.10	-255.46	*-247.48*
#36	-199.19(10)	-214.12	-208.62	-210.44	*-202.41*
#37	-143.12(10)	-198.90	-184.32	-175.14	*-167.87*
#38	-255.91(9)	-277.06	-272.52	*-262.44*	-266.78
#39	-254.15(10)	-272.34	-264.86	-258.69	*-254.18*
#40	-258.54(10)	-277.03	-268.69	-265.25	*-259.90*
#41	-227.15(10)	-260.87	-259.40	-234.67	*-227.15*
#42	-254.17(10)	-276.25	-265.79	-264.32	*-259.35*
#43	-271.58(10)	-291.47	-285.22	-276.09	*-272.63*
#44	-269.95(10)	-305.95	-304.80	-302.42	*-297.55*
#45	-273.51(10)	-296.21	-290.57	-286.74	*-275.13*
#46	-156.97(10)	-188.60	-172.56	-185.82	*-162.94*
#47	-206.90(10)	-257.51	-249.40	*-209.99*	-213.28
#48	-269.61(8)	-280.7	-269.69	-276.23	*-272.03*
#49	-212.30(10)	-256.22	-222.37	-216.33	*-212.51*
#50	-220.06(10)	-255.13	-238.19	-229.54	*-221.73*

Table 8.11. Experimental results (#51 — #100): lp_n

Word	GA-HMM	Re-estimation (7 states)	Re-estimation (8 states)	Re-estimation (9 states)	Re-estimation (10 states)
#51	-182.54(10)	-229.71	-206.85	-204.17	*-182.54*
#52	-191.34(9)	-230.09	-214.92	*-191.34*	-193.99
#53	-176.04(10)	-216.78	-206.00	-190.04	*-176.65*
#54	-184.30(10)	-226.45	-220.94	-207.70	*-190.46*
#55	-161.44(9)	-187.51	-195.05	*-162.07*	-167.17
#56	-167.05(9)	-196.37	-188.57	*-168.27*	-174.67
#57	-240.95(10)	-266.11	-263.71	-244.06	*-240.95*
#58	-136.38(10)	-194.48	-181.13	-163.99	*-140.71*
#59	-241.06(10)	-261.73	-265.53	-246.73	*-243.00*
#60	-232.14(10)	-276.11	-257.76	-252.78	*-233.64*
#61	-192.88(10)	-246.67	-229.13	-221.74	*-193.50*
#62	-163.84(10)	-200.71	-184.78	-187.29	*-174.54*
#63	-233.21(10)	-270.53	-253.37	-248.02	*-233.21*
#64	-173.94(10)	-220.57	-205.14	-185.68	*-181.15*
#65	-215.42(9)	-252.45	-232.60	*-219.75*	-223.91
#66	-182.02(10)	-226.26	-218.93	-222.29	*-186.11*
#67	-208.15(10)	-245.31	-227.65	-228.48	*-208.88*
#68	-190.57(9)	-219.38	-225.37	*-192.26*	-192.47
#69	-195.91(10)	-222.69	-202.27	-206.34	*-200.87*
#70	-81.23(9)	-131.84	-118.68	*-96.63*	-96.98
#71	-248.03(10)	-265.34	-251.87	-258.66	*-249.41*
#72	-258.67(9)	-273.09	-274.21	*-259.27*	-265.92
#73	-193.80(10)	-246.73	-227.69	-215.78	*-198.20*
#74	-252.94(10)	-295.30	-270.83	-295.30	*-252.96*
#75	-238.68(9)	-260.50	-257.11	-239.25	*-238.73*
#76	-208.87(10)	-233.84	-215.64	-225.11	*-209.22*
#77	-165.75(10)	-198.18	-181.40	-167.05	*-165.75*
#78	-206.75(10)	-235.49	-225.32	-220.75	*-208.34*
#79	-267.59(10)	-285.87	-277.51	-270.63	*-269.87*
#80	-200.09(9)	-221.88	-209.26	*-200.30*	-203.70
#81	-250.42(10)	-283.15	-273.60	-263.34	*-252.76*
#82	-238.11(10)	-259.60	-254.84	-246.06	*-239.87*
#83	-180.63(10)	-230.90	-208.50	-198.96	*-180.63*
#84	-263.18(10)	-274.40	-277.23	-272.25	*-263.79*
#85	-267.63(10)	-288.10	-278.40	-273.85	*-270.00*
#86	-164.97(10)	-236.00	-205.71	-201.73	*-165.37*
#87	-209.09(10)	-259.52	-253.31	-228.51	*-211.98*
#88	-107.66(10)	130.67	-139.07	-120.75	*-109.47*
#89	-211.03(10)	-263.15	-239.19	-225.74	*-221.03*
#90	-219.45(10)	-250.68	-263.21	-230.80	*-221.03*
#91	-108.93(10)	-153.84	-151.92	-125.66	*-109.96*
#92	-217.55(10)	-246.00	-236.98	-222.56	*-217.68*
#93	-240.88(10)	-306.68	-295.53	-257.32	*-240.88*
#94	-207.73(9)	-234.19	-214.69	*-207.13*	-209.64
#95	-232.82(10)	-247.42	-247.12	-242.46	*-234.08*
#96	-154.32(10)	-230.33	-210.33	-198.02	*-154.32*
#97	-127.73(10)	-173.79	-150.56	-143.47	*-131.29*
#98	-171.37(10)	-215.31	-214.53	-218.69	*-172.56*
#99	-218.80(10)	-256.15	-237.78	-227.02	*-222.25*
#100	-230.60(10)	-257.67	-258.37	-251.21	*-234.05*

238   8. Genetic Algorithms in Speech Recognition Systems

contribution to the GA-HMM training procedures.

For further illustration, Fig. 8.20 shows a case where the chromosomes have a number of states that are unevenly allocated at the beginning. It can be seen that the chromosomes with 10 states are randomly set to three times more than the number of the chromosomes with 8 states. In spite of this handicap at the beginning, the chromosomes with 8 states began to increase their occurrence and gradually to dominate the entire population. Therefore GA-HMM is considered to be a fairly robust method for speech recognition.

To summarize the whole GA-HMM development, we can conclude with the remarks as follows:

- The use of GA based HMM training procedure, the number of states and the model parameters can be optimized simultaneously;
- The state mutation operator is found to be a robust design for identifying the appropriate number of states of HMM even when uneven initialization has taken place;
- For the implementation of the state mutation operator, GA is a very easy-to-use optimization algorithm. By the use of a very simple genetic operator and state mutation, the appropriate number of states can be found in the word model both efficiently and without any specific knowledge;
- Judging from experimental results, GA-HMM has a better training performance as compared with the Baum-Welch algorithm; and
- Although GA-HMM training may require a longer training time than the Baum-Welch algorithm, this is normally a non-issue problem as the training process is generally off-line operated.

8.6 A Multiprocessor System for Parallel Genetic Algorithms

Despite the application of hybrid-GA formulation for speeding up the pool convergence, the stringent requirement for a training model for both GA-DTW and GA-HMM remains unchanged. There is also the consideration of genetic computational constraint. The fact is that a single processor only allows a single population pool for genetic operations. Considering the magnitude of the feature vectors, this is usually very large in size, and the pool to accommodate the enormous number of chromosomes for an effective genetic operation is doubtful. In the end, this problem leads to pre-mature convergence which yields sub-optimal solutions.

Given that with the Holland's schema theory [Holland, 1975], where GA is said to be quite capable of manipulating a large number of schema in a parallel

8.6 A Multiprocessor System for Parallel Genetic Algorithms   239

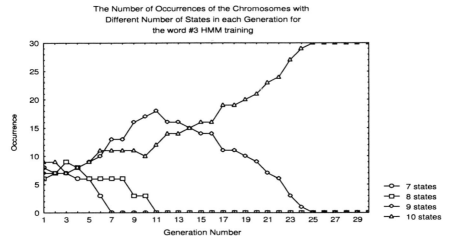

Fig. 8.17. The number of occurrences of the chromosomes with a different number of states in each generation for the word #3 HMM training (trained optimal number of states in the word model is 10)

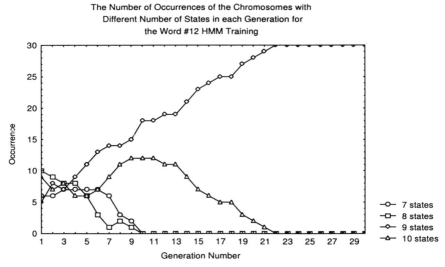

Fig. 8.18. The number of occurrences of the chromosomes with a different number of states in each generation for the word #12 HMM training (trained optimal number of states in the word model is 9)

240   8. Genetic Algorithms in Speech Recognition Systems

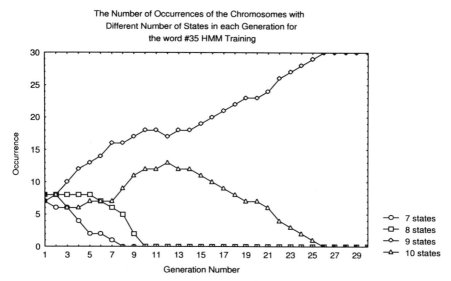

Fig. 8.19. The number of occurrences of the chromosomes with a different number of states in each generation for the word #35 HMM training (trained optimal number of states in the word model is 9)

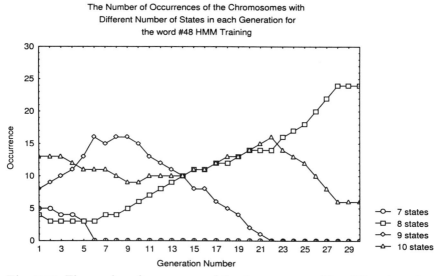

Fig. 8.20. The number of occurrences of the chromosomes with a different number of states in each generation for the word #48 HMM training (trained optimal number of states in the word model is 8)

8.6 A Multiprocessor System for Parallel Genetic Algorithms

fashion, the straightforward way of improving the speed of computation would then be the way of connecting a number processing nodes with a dedicated communication link between them.

This concept has been well reported in Sect. 3.1, where three different types of GA parallel architectures; namely Global GA, Migration GA, and Diffusion GA have been generally described. For the case of speech recognition, which is similar to most of the other GA problems, the bottleneck of computation is not hinged on the processing of genetic operations such as selection, recombination, mutation, fitness assignments and so on. In fact, these operations require little time to complete. What seems to be time consuming, is the evaluation of objective functions. This is particularly true when the functions are nonlinear, constrained and discontinuous.

In these circumstance, neither the Migration GA nor the Diffusion GA is recommended as the architecture for both GA-DTW and GA-HMM computation. The only amenable multiprocessor system would then be the use of Global GA topology. In the following section, a dedicated computing architecture based on the concept of Global parallel GA is developed for speech recognition systems [29].

8.6.1 Implementation

Hardware Requirements. The development of multiprocessor systems is largely based on the structure of the speech recognition system as depicted in Fig. 8.2. It can be seen that the feature extraction is the most important part of such a system. Therefore, a dedicated node for feature extraction is desirable as the front-end processor while the other processing nodes are used for performing the GA evolution.

The front-end processor must be able to capture speech signals at a Nyquist rate. This is usually in the range of 8 KHz to 16 KHz. Ceptrum coefficients are the feature vectors in this system and can be defined as the inverse Fourier transform of the short-time logarithmic amplitude spectrum [158].

In this case, a sampling rate of 16 KHz for cepstrum coefficients is adopted. A 256-point window is used to minimize the signal discontinuities at the beginning and end of each frame [173]. To ensure a continuous real-time operation, the feature extraction process must perform a given time frame within a 16ms time limit. As for the steps needed to calculate the pitch period and a spectral envelope, Fig. 8.21 [159] illustrates the required procedures.

242   8. Genetic Algorithms in Speech Recognition Systems

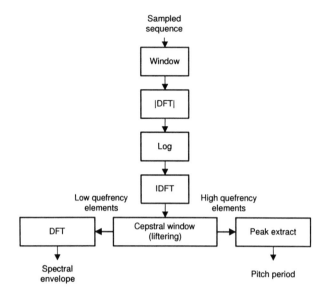

Fig. 8.21. Block diagram for ceptral analysis

To satisfy the required computation, the number of the multiplying operations that must be fulfilled for the performance of windowing, discrete FFT, logarithmic and the inverse discrete FFT are 256, 8192, 2048 and 4096 respectively. This is equivalent of approximately 13,500 multiplications.

These multiplications are very suitable for modern RISC processors or digital signal processors (DSP), while the other operations such as memory access and addition can also be applied. However, depending on the architecture used, about 10% to 100% of the operations cannot be used for executing the parallel multiplication operations. For this particular application, we assume a 25% redundancy rate. Therefore, approximately (13,500 + 4,400 = 17,900) multiplication operations for the feature extraction in every 16ms time frame are required. In term of microseconds per multiplication, this is around 0.89ms (16ms/17,900) per multiplication.

Selection of the node processor. The required computational power mentioned in the previous section can be easily acquired by using modern RISC processors such as the ALPHA from Digital Equipment Corp., the UltraSparc from Sun Microsystems Inc., the POWER system from Motorola Inc, and so on. As these are general-purpose types, which include memory management units with embedded large on-chip first-level caches, they are generally expensive. In addition, the heavy I/O demand of the real-time speech analysis which also requires dedicated high-speed data transfer hardware to interface with the complexity I/O interface, means that RISC processors are not

8.6 A Multiprocessor System for Parallel Genetic Algorithms

practical.

On the other hand, the DSP type is preferred. DSP is a special purpose processor equipped with dedicated hardware to handle the operations which are commonly used for digital signal processing applications. Thus, it has no on-chip cache in the memory management unit.

In fact, DSP can accommodate special architectures. These include the parallel executable multi-bus and the separation of functional units which allow the multiplication, addition, and memory access to be more accessible. Furthermore, the special address mode and instructions that also exist to improve the performance of processing. For example, the bit-reverse address mode for calculating the FFT does not require any additional memory address. The additional employment of a simple I/O interface, fast response interrupt mode and on-chip DMA unit, etc. will also alleviate the burden of heavy I/O demand in real-time processing of the speech.

Having made the necessary comparison in terms of features and functionalities between RISC and DSP processors, it is not too difficult to select DSP processors as the option for the construction of the multiprocessor for this particular application. We have considered several commercial DSP chips as the backbone of the computing architecture, namely:

- TMS320C25 from Texas Instrument Inc.
- TMS320C30 from Texas Instrument Inc.
- DSP56001A from Motorola Inc.
- DSP96002 from Motorola Inc.

A brief account of the analysis of these items was made on the basis of specifications provided by a data sheet supplied by the manufacturer of each DSP. The comparisons are shown in Table 8.12.

TMS320C25 offers the advantages of low cost, simple system design, and excellent availability. However, the 16 bits fixed point of default data type and the 10 MIPS operational speed seriously restricted it to low-end applications, especially as the 16 bits word length was barely able to meet our requirement of high quality speech processing which requires speech sampling with 16 bits resolution. As a node processor, these features could also seriously affect future development for multiprocessor systems.

DSP56001A is another fixed point DSP that has all the functionalities possessed by the TMS320C25, but it has a longer word length - 24 bits fixed point and faster instruction execution. It does fulfill the purposes of our speech front-end processor. However, in the application of GA-DTW and GA-HMM, where the distortion measurement between the two speech-patterns

Table 8.12. Comparison of four DSP chips: TMS320C25, DSP56001A, DSP96002, and TMS320C30

Criterion	TMS320C25	DSP56001A	DSP96002	TMS320C30
Data type	16 bits fixed point	24 bits fixed point	32 bits floating point	32 bits floating point
Speed (MIPS)*	10‡	16.67‡	16.67‡	16.67‡
Complexity†	1	2	3	3
Availability	Excellent	Good	Good	Excellent
Power Dissipation*	0.76W	0.97W	1.75W	3.2W

Note:
* — 33.33 MHz clock speed version.
† — Relative system-design complexity used TMS320C25 as reference.
‡ — Execute more than one operation in a single instruction
    TMS320C25 - 2 operations per instruction
    DSP56001A - 4 operations per instruction
    DSP96002 - 5 operations per instruction
    TMS320C30 - 3 operations per instruction

in GTW is unbounded, these contradicted the default data representation of DSP56001A whose range (-1.0 to 1.0) is bounded. As for the case of GA-HMM, the log probabilities can be very small and cannot be represented in 24 bits fixed point representation.

AS for DSP96002 and the TMS320C30 (C30), both of these are 32 bits floating point DSPs as well as having a larger dynamic range for data handling. This feature is well suited for the application of GA-DTW and GA-HMM. The instruction cycle is also fast. The additional functions such as dual external buses, on-chip DMA controller, and synchronous serial ports to deal with I/O demanded applications are handy for both software and hardware development.

Considering that both the DSP96002 and the TMS320C30 (C30) are acceptable for the computing architecture development, the difference in choice between the two was small and restricted to only one criterion which was cost. Despite the fact that the DSP96002A may offer extremely high performance in that it can execute five operations in one instruction, it is also much more expensive than the C30. But C30 had the essential advantage over DSP96002A in the development phase which was its availability. Based on the above arguments, it was not difficult to conclude that C30 is a better choice as the core processors in our multiprocessor system.

Memory configuration. Memory configuration is another important factor affecting system performances in the design of a multiprocessors system. Generally, there are three types of memory in a multiprocessor system: local memory, global memory, and bootstrap memory.

8.6 A Multiprocessor System for Parallel Genetic Algorithms

Local memory is used to store local data and program codes for execution within each processing node. Global memory enables inter-process communication while Bootstrap memory stores the program codes for the power-up procedures using Read Only Memory (ROM). Since the target application of our multiprocessor system was speech processing in the context of GA applications, then a 64k words (each word has 32 bits) local memory per processing node was required. Should the flexibility need to be increased, the processing node had the spare capacity to increase the local memory up to 128k words. Fast static memory was also required for local memory to access the memory of C30 at the highest speed. The inter-process communication (IPC) channel was achieved by the 8k words global memory, while its communication overhead was alleviated by implementing a fast static memory.

The next step was to configure the memory blocks of the system. Obviously, local memory is located in its own processing node. Since the global memory is designed as a shared resource for all the processing nodes in the multiprocessor system, a separate global memory board had to be installed, and the processing nodes to access the global memory via a global bus.

The global memory board was also equipped with an arbitrator to ensure that only one processing node could access the global memory at a time. Due to arbitration delay and the driving buffer delay of the global bus, the access time for global memory was longer than the local memory. In this case, the dual external buses of the processor C30 came in to synchronize the different memory access speeds.

The dual external buses of C30 had two external buses: the primary bus and the expansion bus. Primary bus is the main bus of C30 that provides full speed to access main memory of a computer system. Obviously, the local memory is connected to the primary bus in order to obtain full computational power. The expansion bus of C30 is used to connect with the I/O devices. It usually has a slower access speed than accessing the main memory, thus, an expansion bus that connects to the global bus is preferred.

Bootstrap memory also connects with the primary memory of C30 in each processing node. The bootstrap memory is stored the program code for the power-up procedure and the application code. Due to the slow access speed of the bootstrap memory, C30 will not directly execute the application's program code in the memory. Instead, the code will first be downloaded to the local memory by the power-up procedure and then C30 will execute the code in the local memory at full speed. The overall memory configuration is depicted in the following figure.

246   8. Genetic Algorithms in Speech Recognition Systems

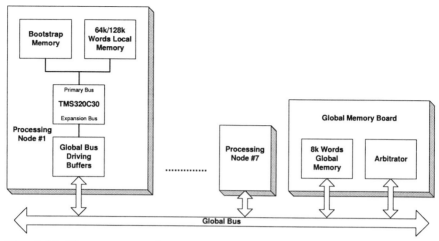

Fig. 8.22. The memory configuration of the multiprocessor system

Processor Synchronization. Processor synchronization is another important factor to be considered in any multiprocessing system design. This can be demonstrated by the example as follows:

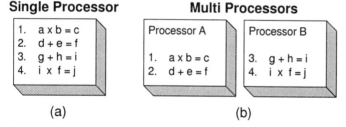

Fig. 8.23. Diagram to illustrate the importance of processor synchronization. (a) Codes executed in a single processor. (b) Codes executed in two processors

According to the formulation as shown in Fig. 8.23(a), the program code is executed sequentially on a single processor. Fig. 8.23(b) shows the same program code that can execute in parallel with two processors. In Fig. 8.23(b), the instructions 1, 2, and 3 can be executed in any arbitrary order. There are no data dependencies among these instructions. However, the computation of instruction 4 is dependent upon the results of instructions 2 and 3. As instructions 2 and 4 are executed in parallel by processors A and B, a synchronization mechanism is therefore desirable in order to ensure that the processor A executes instruction 2 before processor B executes instruction 4. This mechanism is essential and is known as processor synchronization in the design of multi-processor system.

The simplest way to implement processor synchronization is to reserve some memory space in the global memory and use this as the synchronization registers. As indicated in Fig. 8.23(b), the synchronization can be achieved by placing an active flag in the synchronization register when processor A finishes the execution of instruction 2. Processor B executes instruction 4 only after the flag in the status register is set or being active. Thus, the synchronization among processors is achieved via monitoring the status of the synchronization register.

This above method of synchronization is simple and easy to implement. However, the use of polling the active status of the synchronization register induces extra communication overheads to the system. Therefore, instead of polling, the function of interrupt should be used between the processing nodes. The synchronization uses four external interrupt request lines in the C30; namely /INT0, /INT1, /INT2, and /INT3 as the synchronizing signal lines. /INT3 is the global synchronizing signal line which ensures that all /INT3 of the processing nodes in the multiprocessor system are connected together in the global bus and the /INT0, /INT1, and /INT2 in each processing node are the three separated synchronizing signal lines. Each of these three lines of processing node are interconnected independently in the global bus, so that each processing node can synchronize any other three other processing nodes in the same manner. The prototype system is shown by the photograph in Fig. 8.24.

Global Bus. In order to connect all of the processing nodes and the global memory board, a back-plane bus is implemented as shown in Fig. 8.25. The processing nodes are connected to the global bus via two DB96 connectors. A total of seven processing nodes and one global memory board are connected via a global bus. This amount of nodes may limit the driving-buffers capability of the global bus, but at the same time, any higher number of nodes could also cause deterioration of the signal's quality in the global bus, which in turn could cause erroneous data transfer. Figure 8.26 is a photograph of the multiprocessor system with four processing nodes and one global memory board. It is now a system can be used for our parallel GA applications.

8.7 Global GA for Parallel GA-DTW and PGA-HMM

Having established the essential hardware requirements, the implementation of parallel GA-DTW and GA-HMM can proceed. For both applications, the partition of the software and the method of communication between each processing module are the major considerations for implementation. In the Global GA architecture, these problems are closely related to the topology

248   8. Genetic Algorithms in Speech Recognition Systems

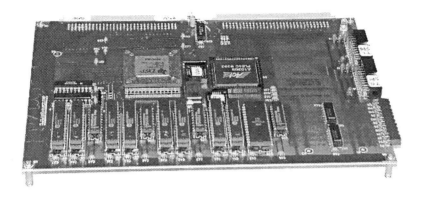

Fig. 8.24. The photograph of the processing node of the multiprocessor system

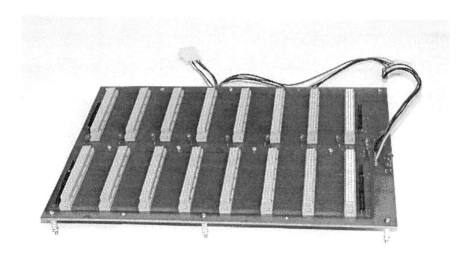

Fig. 8.25. The photograph of the global bus of the multiprocessor system

8.7 Global GA for Parallel GA-DTW and PGA-HMM

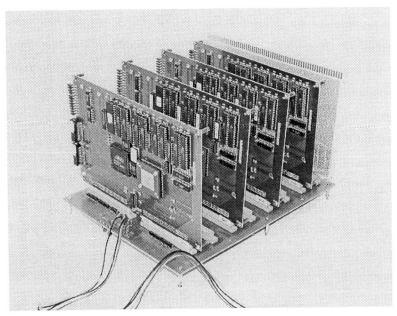

Fig. 8.26. Photograph of the multiprocessor system

of the multiprocessing system.

A schematic diagram of this GA processing system is shown in Fig. 8.27. The global GA divides a large population into smaller sub-populations and each sub-population is stored in each node of the multiprocessor system. In each node, the node processor executes the GA evolution cycle on its own sub-population separately. Therefore, the more nodes in the system, the smaller number of individuals that have to be processed in the sub-population pool and hence a shorter time is required to compute each generation.

Obviously, the GA cycle in the Fig. 8.27 should be different from the traditional one. In each node, it manipulates its own sub-population only. In order to obtain the global statistics of the entire population, a communication operator called the migration operator is designed and added to the GA cycle.

In a migration process, each sub-population selects a number of better and fitter individuals (chromosomes) as migrants. This strategy ensures that the fitter individuals will have a higher chance to survive in the subsequent generations and hopefully, more offspring can be mutated amongst sub-populations so that premature convergence is avoided.

This operator consists of two modes of operation as demonstrated in Fig. 8.28. In the first mode, the selected individuals are transferred to the

250   8. Genetic Algorithms in Speech Recognition Systems

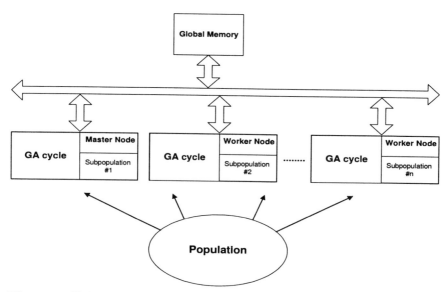

Fig. 8.27. Global GA implementation in the proposed multiprocessing platform; where $n = 2, \cdots, 7$

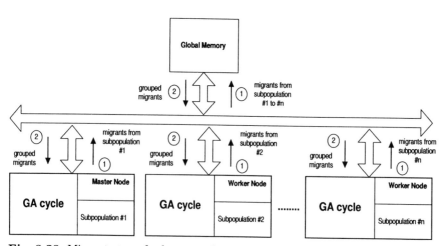

Fig. 8.28. Migrants transfer between the nodes

global memory. In the meantime, the master node will act as the coordinator to synchronize the migration operator in each node. The process of migration is completed as long as the master node signifies a warning signal to each worker node.

In the second phase of migration, the worker node will replace the chromosomes in its own population pool upon receiving migrants from the Master node. Then, a new sub-population is reproduced for the next GA evolution cycle. A clearer schematic diagram of this migration mechanism is shown in Fig. 8.29.

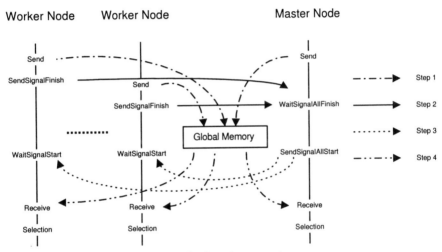

Fig. 8.29. The communications of migration operator

It is only this concept of the migration, which enables each sub-population to ensure a well mixed population to form an entirely statistical pool in each generation cycle. Without this essential procedure, the parallel versions of GA-DTW and GA-HMM cannot be realized in a satisfactory manner.

8.7.1 Experimental Results of Nonlinear Time-Normalization by the Parallel GA-DTW

To verify both software and hardware development of the proposed parallel architectures, the parallel GA-DTW is the first to be executed on both single and two processing nodes systems.

Consider a database of 90 different isolated English words which were extracted from the TIMIT speech database and split into two sets of data.

252   8. Genetic Algorithms in Speech Recognition Systems

Set 1 has 70 utterances for 10 of the words, and set 2 has one utterance for another 80 words. The sampling rate is 16,000 Hz. and the endpoint detection of each utterance is labeled by the manufacturer of the TIMIT database. A set of 12-order cepstral coefficients is used as the feature vector.

The sub-population size is limited to 40 chromosomes; the crossover rate is 0.6 and the mutation rate is 0.03333. The evolutionary cycle terminates at the end of the *40*-th generation.

For each of the words in set 1, an arbitrarily chosen utterance was selected as the "test template" and used to match with the remaining utterances of the same words in the set 1 database. In addition, the "test template" was also utilized for the matching of the other 80 utterances in set 2. The measures of the matching performance for the two experiments are defined as follows:

- M_s = the average of the distances between the "test template" and the same word in the remaining 69 utterances in set 1;
- δ_s = the standard deviation of the distances between the "test template" and the same word in the remaining 69 utterances in set 1;
- M_d = the average of the distances between the "test template" and those utterances of 80 different words in set 2; and
- δ_d = the standard deviation of the distances between the "test template" and those utterances of 80 different words in set 2.

In the experiments, each warping path had five relaxed beginning and ending points, i.e., $\Delta x = 5$ and $\Delta y = 5$. The local continuity constraint was of Type IV. The allowable region was defined as $Q_{max} = 3$ and $Q_{min} = 1/3$.

Table 8.13. Experimental results of the PGTW by using one processing node

word	One processing node			
	M_s	δ_s	M_d	δ_d
1	8.14	2.66	15.82	40.22
2	10.04	11.29	22.74	51.03
3	8.19	4.03	17.46	41.11
4	5.77	8.06	17.02	42.74
5	11.02	0.53	12.37	39.26
6	6.82	3.82	20.56	71.05
7	13.90	39.90	19.78	45.94
8	3.62	3.56	25.85	90.30
9	2.35	1.21	12.77	40.06
10	2.67	0.60	15.58	47.15

The results obtained from parallel GA-DTW are tabulated in both Tables 8.13 and 8.14 for the operations of one and two processing nodes respectively. Judging from the Ms performance,which is shown in Fig. 8.30, the parallel GA-DTW using two processing nodes induces smaller Ms values

8.7 Global GA for Parallel GA-DTW and PGA-HMM

Table 8.14. Experimental results of the PGTW by using two processing nodes

word	Two processing nodes			
	M_s	δ_s	M_d	δ_d
1	7.16	2.23	14.65	37.54
2	8.31	15.27	21.36	46.26
3	7.97	3.68	16.41	39.37
4	4.93	4.46	15.72	43.16
5	9.46	0.44	11.82	38.28
6	6.05	3.42	19.50	69.86
7	12.32	35.59	18.75	47.47
8	3.18	2.86	24.13	86.48
9	2.15	1.14	12.01	39.09
10	2.31	0.61	14.90	44.38

than those obtained from the single processing node.

For the same number of terminating generations, the results generated by two processing nodes is better. This is largely due to the fact that it has a total of 80 chromosomes which is double the size of the single processing node. This also illustrates that the more chromosomes there are used in GA operations, more potential solutions become available, which also increases the searching space for the global solution.

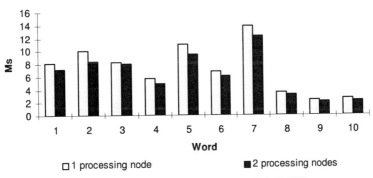

Fig. 8.30. The experimental results of parallel GA-DTW

Table 8.15 shows the execution time used in the parallel GA-DTW with one and two processing nodes. Although the number of chromosomes explored in the second experiment is double the number of chromosomes in the case of single processor, the execution times of the second experiment was found to be only slightly larger. This indicates that the parallel GA-DTW is very

Table 8.15. Execution time (in second) of the parallel GA-DTW

Word	one processing node	two processing nodes
1	5.0	5.1
2	4.7	4.8
3	6.2	6.4
4	4.2	4.3
5	3.8	3.9
6	4.6	4.8
7	3.2	3.3
8	5.8	5.9
9	5.1	5.2
10	6.1	6.3

effective when its comes to execution time.

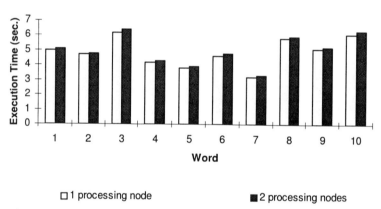

Fig. 8.31. The chart of the execution time of PGTW by using one and two processing nodes

Experimental Results of HMM Training by Parallel GA-HMM. To verify the parallel GA-HMM, four sets of experiments were performed. The first experiment was for one processing node, the second for two processing nodes, and so on.

The testing data as well as the control parameters used for the four experiments were the same as those for parallel GA-DTW. A set of 10 words testing data with 10 utterances for each word was adopted. The utterances were framed with 320 samples, and each frame was analyzed by cepstral analysis, while the speech feature of the frame required ten-order cepstral

8.7 Global GA for Parallel GA-DTW and PGA-HMM   255

coefficients.

In each experiment, ten HMM of the ten words of the testing data were trained and each HMM was also trained from its associated ten training observation sequences. A uniform random number generator was required to generate the initial model parameters so that these could be normalized and at the same time satisfy the constraints mentioned in Eqns. (8.16 and 8.17). The GA parameters used for the experiments were as follows:

Table 8.16. Control parameters for GA-HMM

Population size	40
Crossover rate	0.9
Mutation rate	0.01

The average log probability p_{same} of the HMM generated by the ten training observation sequences of the same HMM were computed. All of the experiments terminated after 20,000 generations. The results of the four experiments are listed in Table 8.17. Judging from the size of p_{same}, the more processing nodes used, the better the quality of the HMM model. The result in word 3 of the third experiment is an exceptional case in that its p_{same} is smaller than the p_{same} of the word 3 in the second experiment. This is because the random search property of GA has no better solution when the 20000-th generation is reached.

Table 8.17. Experimental results of the parallel GA-HMM

Word	# of processing nodes			
	one	two	three	four
1	-12.685	-12.254	-11.784	-11.269
2	-13.397	-13.054	-12.410	-12.016
3	-12.461	-11.641	-11.856	-11.433
4	-12.715	-12.018	-11.873	-11.316
5	-12.417	-11.879	-10.960	-10.152
6	-12.713	-12.195	-11.452	-10.510
7	-12.516	-11.763	-10.930	-10.539
8	-12.478	-11.937	-11.029	-10.096
9	-13.412	-13.389	-12.585	-11.830
10	-13.374	-13.126	-12.227	-11.462

The execution time used for the parallel GA-HMM with one to four processing nodes are listed in Table 8.18 . Judging from Fig. 8.33, the execution time used for those experiments also increases slightly when the number of the processing node increases.

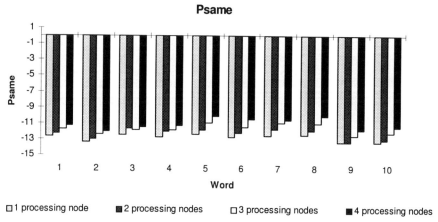

Fig. 8.32. The chart of p_{same} of the parallel GA-HMM

Table 8.18. Execution time (in second) of the parallel GA-HMM

Word	# of processing nodes			
	one	two	three	four
1	2430	2494	2533	2561
2	2312	2371	2414	2440
3	2462	2525	2565	2596
4	2212	2270	2309	2338
5	2802	2864	2902	2934
6	2539	2600	2644	2678
7	2451	2507	2544	2574
8	2673	2731	2775	2805
9	2434	2494	2529	2557
10	2652	2716	2756	2784

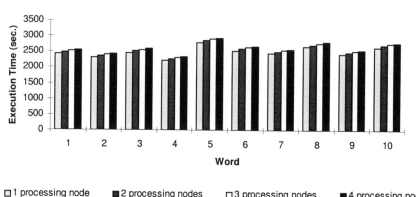

Fig. 8.33. The chart of the execution time of the parallel GA-HMM

8.8 Summary

We summarize this chapter as follows:

1. Two major model training methods based on GA have been proposed for speech recognition, namely GA-DTW and GA-HMM. In the GA-DTW experiments, the results have shown that a better quality model was found as compared with the DTW algorithm alone. By the use of the measure $|M_s - M_d|$, the decision making logic can be rendered more reliable in recognizing both similar and dissimilar utterances. For non-trivial cases, the finding of K-best paths in GA-DTW is natural and simply by taking the K-best fitness values;
2. Experimental results also indicate that the training model obtained using the GA-HMM method is better than the HMMs trained by the Baum-Wlech algorithms, see Tables 8.10 and 8.11. The capability of parallel GA-HMM for searching the best model is very high as the training process can escape from the initial guess. In addition, the function of parents re-distribution in the master processor is an added advantage; and
3. It should be noted that the use of parallel GA for speech recognition is rather different from the one used for active noise control in Sect. 5.3. The emphasis for speech recognition is on obtaining the best possible, if not an optimal model, as the template reference model in the speech database. Therefore, the execution time in the off-line training exercise is a non-issue item.

9. Genetic Algorithms in Production Planning and Scheduling Problems

In any manufacturing system, an effective production planning and scheduling programme is always desirable. Such a scheme involves the power to solve several mathematical intangible equations. Ad hoc solutions may be obtained but often fail to address various involved issues.

In this chapter, GA is used to optimize a number of functions to improve the effectiveness of the production planning and scheduling schemes. By the use of MOGA, see 3, large-scale multi-type production with the considerations of lot-size and multi-process capacity balance problems can all be solved in a simultaneous manner. In this way, manufacturers can respond to changing market requirements in a timely manner and fulfill the needs of customers. This method is a noted improvement on any existing techniques, and also in practice, provides a new trend for the design of manufacturing systems.

9.1 Background of Manufacturing Systems

In the 1950s, industrial companies were already focusing their attention on the development of production and manufacturing schemes that would enhance their production rates and profits. The most noticeable production and inventory control and management programme (PICM) [178, 96] played a fundamental role in this arena. PICM is an effective control and management strategy for overseeing the total flow of production processes, starting from the acquisition of raw materials to the delivery of finished products for customers. It is an informative scheme that can benefit both managers and customers. In addition, it also facilitates manufacturing resource to satisfy marketing requirements. A schematic diagram of PICM is shown in Fig. 9.1.

By the end of the 1960s, one particular technique called material requirement planning (MRP) had been established in the USA and Europe. MRP serves well as a tool for managing materials, although in the context of capacity, it seems a much less attractive proposition to deal with. This deficiency can be overcome by the use of the closed loop MRP [145] which

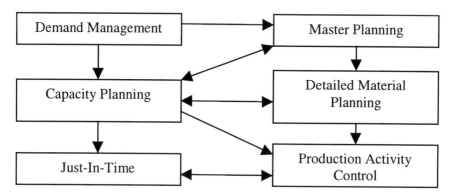

Fig. 9.1. Production and inventory management system

offers a complete solution for the management of both materials and capacity. A detailed plan of this system is depicted in Fig. 9.2.

As society and industry advanced in a synchronized manner, the demand for changing the manufacturing policy to meet customer delivery requirements has become of paramount importance. An ill conceived planning will inevitably add to increased costs. The desire for a management system with the ability to address the control of materials, capacity requirements, cash flows and cost minimization is now commonly expected. Such a system is an improved version of the closed loop MRP and has hence been renamed manufacturing resource planning (MRPII). This system became very popular in the 1970s and 1980s as computer technology developed rapidly.

In many ways, MRPII still falls short of being a complete system. In order to reach a solution whereby irregular demand and purchasing pattern of a manufacturer's co-operating suppliers can be smoothed out may involve the implementation of very complicated management schemes. Furthermore, MRPII makes oversimplistic assumptions, and tends to be nervous of the scheduling and multiplier effect. Therefore, its ability is limited to the design of flow-based, batch job and specific industrial operations [100, 151].

In addition to carrying out MRPII development, the concept of Just-in-Time (JIT) manufacturing was initiated in Japan. This provides a tight manufacturing management schedule whose philosophy is to ensure *"the right items of the right quality and quantity in the right place at the right time"* [31].

A proper use of the JIT principle in manufacturing can bring about benefits in the area of waste reduction, increased ability, productivity and

9.1 Background of Manufacturing Systems   261

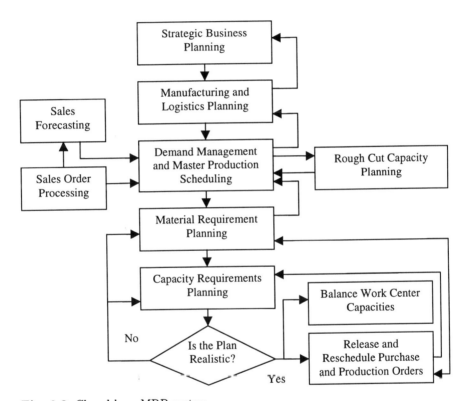

Fig. 9.2. Closed loop MRP system

efficiency, communications reliability, higher profits and customers satisfaction, etc. It has emerged as a means of obtaining the highest levels of usage out of limited available resources. For these reasons, JIT has become very popular. However, it too has drawbacks. For example, cultural encumbrances (such as employee involvement and participation), decreased safety stocks, loss of autonomy and resistance to change can occur [206, 191, 186, 111].

Having realized both the advantages and disadvantages of MRPII and JIT, a merge of the two techniques was then proposed [190, 104, 105, 61, 227]. Embedding JIT into MRPII based on back-flushing and phantom features was proposed [61], and an optimal push-and-pull control strategy also provided an effective integration of the two methods [190, 104, 105].

However, the initial integration of MRPII with JIT was only applied to the production control level, and the production planning level was seldom considered. This problem was not solved until the development of an earliness/tardiness production planning and scheduling system (ETPSP) was established [199, 12]. This method fits into the JIT philosophy perfectly, whereby the production costs due to an early production schedule or late delivery can be under controlled.

All the above techniques discussed can only solve production planning and control problems in a piecemeal manner. A much more effective combination would be the forming of a unique but central scheme for controlling production management. Such a system could play a critical role in coordinating and integrating operational decisions, including marketing, engineering, finance, manufacturing and human resource [73]. By utilizing available capacity, a medium-term production planning process to meet changing market requirements can also be realized. In addition, it may provide a support for other procedures and plannings within a manufacturing system. This includes business planning, material requirement planning, purchase planning, capacity planning and final assembly scheduling, etc. Such a system is now called the Master Production Planning and Scheduling (MPSP) system. The overall structure of an MPSP is shown in Fig. 9.3.

The key issue on which the success of MPSP hinges is the process of minimization. The conventional MPSP takes the minimization of total production cost, or the maximization of production output as the objective. In the sense of JIT, this is a measure that can result in dangerous consequences, particularly in a changing and competitive market. A more amenable alternation is to use the due date criterion [52]. Its timely ability to adjust to ever changing requirements is more favorable to manufacturers. Hence, a concise MPSP is largely dependent upon the utilization of solutions

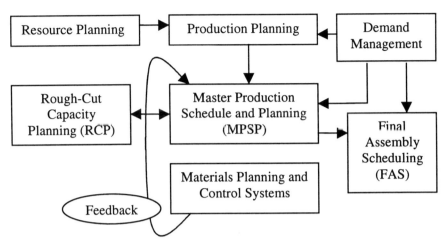

Fig. 9.3. Overall MPSP system

produced by ETPSP.

However, despite a number of designs having been proposed to address the ETPSP problems [94, 95, 44, 211, 212, 2, 124, 226, 97], their applications are still restricted to constant capacity. Considering that the nature of ETPSP is one in which the optimization functions are basically nonlinear and discrete, the current techniques have failed to provide an efficient solution other than the incidental one-product or at the most, two-product cases. In addition, their calculations also neglect the lot-size consideration. In the event of a large scale problem, even the key-process and shrinking-feasible-field methods offer no solutions due to computational difficulties [226, 97, 132, 133].

Having reviewed every possible aspect of the MPSP structure, and these include the emphasis of ETPSP methodology, the use of GA would serves an ideal method, as well as an effective optimization system for both production planning and scheduling. It is therefore the purpose of this chapter to outline its relevant features.

9.2 ETPSP Scheme

In order to merge GA into the domain of ETPSP, the fundamental issues of ETPSP should be clearly brought out. As the main objective is orientated around the need to meet changing market requirements, in which the production cost is largely governed by early or tardy production, the principle of the ETPSP method is to find an optimal or near-optimal MPSP, which can minimize the total cost of penalties due to earliness and tardiness and yet maintain the manufacturing capacities in a MPSP horizon can also be confirmed [52, 94, 95].

Without loss of generality, the notations for an ETPSP problem are stated as follows:

- N: the number of products
- i: the index of products, generally $i = 1, 2, \cdots, N$
- M: the number of processes
- j: the index of processes or assembling stages, generally $j = 1, 2, \cdots, M$
- T: the length of a production scheduling and planning horizon
- k: the index of a planning horizon, generally, $k = 1, 2, \cdots, T$
- *Product i*: the name of the i-th product
- *Process j*: the name of the j-th process or assembling stage
- *Period k*: the name of the k-th period in a production scheduling and planning horizon
- $d_i(k)$: the requirement quantity of Product i in Period k
- $c_j(k)$: the available capacity of Process j in Period k
- w_{ij}: the unit capacity requirement of Product i for Process j. It is assumed that the unit capacity requirements of all products should be kept constant along a horizon
- l_i: the initial inventory quantity of Product i, $l_i < 0$, implies the initial shortage of Product i
- $p_i(k)$: the planning production quantity of Product i in Period k
- α_i: the unit time earliness penalty of Product i
- β_i: the unit time tardiness penalty of Product i, α_i and β_i can be determined by the inventory cost and tardiness compensation in practice, generally $\alpha_i > \beta_i$
- s_i: the production lot-size of Product i

9.2.1 ETPSP Model

One of the main objectives of ETPSP is to optimize a lot-size production schedule in an MPSP horizon, in the sense that the conditions of total cost for earliness and tardiness penalties and the manufacturing capacities are all satisfied. The ETPSP area of concern can be described in the following discrete form:

Problem (P)

$$\min_{P} = \sum_{i=1}^{N}\sum_{k=1}^{T}\left\{\alpha_i\left[l_i + \sum_{t=1}^{k}p_i(t) - \sum_{t=1}^{k}d_i(t)\right]^+ + \beta_i\left[\sum_{t=1}^{k}d_i(t) - \sum_{t=1}^{k}p_i(t) - l_i\right]^+\right\} \quad (9.1)$$

subject to:

$$\sum_{i=1}^{n} w_{ij} p_i(k) \leq c_j(k)$$

$$0 \leq p_i(k) \in S_i, \qquad S_i = \{r \cdot s_i, \ r = 0, 1, \cdots\}$$

where $(r)^+ = \max\{0, x\}$, $i = 1, 2, \cdots, N$; $j = 1, 2, \cdots, M$ and $k = 1, 2, \cdots, T$.

9.2.2 Bottleneck Analysis

For a large-scale ETPSP problem, a bottleneck phenomenon appears when both multi-type production and large numbers of capacity constraints are handled simultaneously. This problem is further illustrated by considering the number of constraints $S(N, M, T)$ placed on the scale of Problem (P) for which the formula takes the form:

$$S(N, M, T) = (M - s_{key-process}) \times T + N \times T \approx (M + N) \times T \qquad (9.2)$$

where $s_{key-process}$ is the number of key-processes. Given that with the information provided from Table 9.1, it can be easily concluded that the function of $S(N, M, T)$ increases with N (the number of products) and M (the number of processes). Should both N and M be large, the optimization fails miserably using conventional methods. On the other hand, such a large-scale ETPSP problem, even with the lot-size consideration and the balancing multi-process capacity presents no difficulty for GA.

Table 9.1. Number of constraints $S(N, M, T)$

	$M = 10$	$M = 20$	$M = 30$	$M = 40$	$M = 50$	$M = 100$	$M = 200$
$N = 1$	110	210	310	410	510	1010	2010
$N = 2$	120	220	320	420	520	1020	2020
$N = 5$	150	250	350	450	550	1050	2050
$N = 10$	200	300	400	500	600	1100	2100
$N = 20$	300	400	500	600	700	1200	2200

Note: Horizon of production scheduling and planning is $T = 10$.

9.2.3 Selection of Key-Processes

For any practical manufacturing system, key-processes exist. An effective MPSP structure should also take this into account. The best way to treat this type of problem is to impose constraints on the production capacity. Should the optimization fulfill these constraints, the non key-process capacities are also satisfied accordingly. Since this issue has its own objective function, the computational process can be made much simpler by selecting the MO formulation. Then, the selecting key-processes problem can be solved by imposing with a upper-capacity-function $L(\alpha,t)$, where $\alpha = [\alpha_1, \alpha_2, \cdots, \alpha_N]^T$. The expression for can be written as:

$$L(\alpha,t): \begin{cases} L_1(\alpha,t): \alpha_{11}r_{11}(t) + \alpha_{21}r_{21}(t) + \cdots \alpha_{N1}r_{N1}(t) \\ L_2(\alpha,t): \alpha_{12}r_{12}(t) + \alpha_{22}r_{22}(t) + \cdots \alpha_{N2}r_{N2}(t) \\ \cdots \\ L'_M(\alpha,t): \alpha_{1M'}r_{1M'}(t) + \alpha_{2M'}r_{2M'}(t) + \cdots \alpha_{NM'}r_{NM'}(t) \end{cases} \quad (9.3)$$

with $t = 1, 2, \cdots, T$ and

$$M' \leq M, \quad \sum_{i=1}^{N} \alpha_{ij} = 1; \quad j = 1, 2, \cdots, M'; \quad \alpha_{ij} \in [0,1] \quad (9.4)$$

where $L_j(\alpha,t)$ is the capacity-function of Process j; $\alpha_{ij} \in [0,1]$ is a capacity-assign ratio and the production quality of Product i produced by Process j in Period k takes the form

$$r_{ij}(k) = \frac{c_j(k)}{w_{ij}}, \quad i = 1, 2, \cdots, N; \; j = 1, 2, \cdots, M; \; k = 1, 2, \cdots, T \quad (9.5)$$

whereas the processes whose capacity-functions just partly coincide with the upper- capacity-function $L(\alpha,t)$ are termed as the key-processes[132, 133].

9.3 Chromosome Configuration

In the context of applying GA into ETPSP formulation, an appropriate chromosome structure would greatly enhance the computation process. In this particular application, a real- number representation is adopted instead of the commonly-used binary version [47]. For each chromosome in a real-number representation takes the following sequence:

$$p_1(1)p_1(2)\cdots p_1(T)p_2(1)p_2(2)\cdots p_2(T)\cdots\cdots p_N(1)p_N(2)\cdots p_N(T) \quad (9.6)$$

As an example for illustration based on the information stated in Table 9.2, the chromosome structure should take the associated representation as shown in Fig. 9.4

9.3.1 Operational Parameters for GA Cycles

Similar to other GA applications, the genetic parameters used are unique for each application although the variations may not deviate too far from the norm. The following items are dedicated to this particular application in detail.

| 10 | 5 | 15 | 25 | 30 | 5 | 60 | 20 | 40 | 50 | 70 | 30 | 24 | 36 | 12 | 72 | 108 | 6 |

Fig. 9.4. Chromosome structure

Table 9.2. An example of production quantity

	Period 1	Period 2	Period 3	Period 4	Period 5	Period 6	Lot-size
Product 1	10	5	15	25	30	5	5
Product 2	60	20	40	50	70	30	10
Product 3	24	36	12	72	108	6	6

– *Initialization:* The population pool is randomly generated with real-number strings to ensure diversity;
– *Parent selection:* A roulette-wheel-selection technique applies. A parent selection procedure operates as follows:
  1. sum the fitness of all chromosomes in the population,
  2. generate "n", a random number between "0" and the total fitness value, and
  3. return the first population chromosome, the sum of whose fitness and that of proceeding population chromosome is greater than or equal to n;
– *Fitness:* This takes the form of a linear normalization. It simply converts the evaluations of chromosomes into fitness values, in order to avoid premature convergence. For example, the evaluation of a 10-bit-long chromosome and the calculation of linear normalization fitness can be demonstrated as listed in Table 9.3.

Table 9.3. Fitness technique – linear normalization

Index	1	2	3	4	5	6	7	8	9	10
Original Evaluation	23	36	680	69	78	45	58	17	89	35
Ordered Index	8	1	10	2	6	7	4	5	9	3
Ordered Evaluation	17	23	35	36	45	58	69	78	89	680
Fitness ($c = 10$, $r = 20$)	10	30	50	70	90	110	130	150	170	190

Note: Fitness f_i starts with a constant initial value $c = 10$, and decreases linearly with decrement rate $r = 20$, namely $F_i = c + (i-1) \times r$, $r = 1, 2, \cdots, 10$.

– *Genetic operators:* A two-point crossover and a real-creep mutation are used as the genetic operators to perform the evaluation;
– *Surviving schemes:* Two surviving schemes, an elitist strategy and a steady-state reproduction are introduced:
  1. elitist strategy is used to fix the potential best number loss by copying the best member of each generation into the succeeding generation, and
  2. steady-state reproduction is used to replaces only one or two individuals at a time rather than all individuals in the population. This process discards the children that are duplicates of current individuals in the population rather than insert them into the population;

- *Genetic parameters:* The adaptive genetic parameters are used to ensure population diversity and to speed up the convergence, i.e.,
  1. the population size $S = 100$
  2. the crossover rate $p_{crossover} = 0.9$
  3. the mutation rate $p_{mutation} \leq 0.1$ and $p_{mutation} = C \cdot (|F_{max} - F_{min}|)^{-1}$, where C is a predefined constant, F_{max} and F_{min} are the maximum and minimum fitness of all chromosomes in one generation, respectively.
- *Termination criteria:* Two termination criteria are introduced:
  1. the same solution may continuously be obtained for ten reproduction trials; and
  2. the maximum generation is larger than the preset trial times which have been stated in advance.

9.4 GA Application for ETPSP

9.4.1 Case 1: Two-product ETPSP

The best way for demonstrating the use of GA for the design of manufacturing management is to consider a simply manufacturing system which consists of a 2-product and 10-assembling-stage in a 12-period MPSP horizon. According to different products, the required information is as follows:

- Number of products $N = 2$;
- Number of processes $M = 10$;
- Length of a planning horizon $T = 12$;
- Earliness and tardiness penalties: $\alpha_1 = 10$, $\alpha_2 = 10$; $\beta_1 = 15$, $\beta_2 = 12$;
- Capacity requirement w_{ij} $(i = 1, 2; j = 1, 2, \cdots, 10)$, see Table 9.4;
- Order quantity $d_i(k)$ $(i = 1, 2; k = 1, 2, \cdots, 12)$, see Table 9.5; and
- Available capacity $c_j(k)$ $(j = 1, 2, \cdots, 10; k = 1, 2, \cdots, 12)$, see Table 9.6.

By the use of the GA approach, the ETPSP with lot-size consideration and multi-process capacity balancing are obtained as tabulated in Table 9.7.

Table 9.4. Capacity requirement w_{ij}

i	\multicolumn{10}{c}{j}									
	1	2	3	4	5	6	7	8	9	10
1	1.0	0.6	0.8	0.3	0.7	1.5	1.2	1.1	0.9	0.4
2	0.6	0.8	1.3	2.0	0.7	2.1	0.6	0.8	0.2	0.1

Note: i is the product type, j is the assembling stage

It can be seen from Table 9.5 and Table 9.7 that the difference between the total order requirements and total production quantities is not more than just one lot-size, i.e., 5 or 10 units. That is,

Table 9.5. Order quantity $d_i(k)$ and total requirement

i	\multicolumn{12}{c}{k}	Total Requirement
	1	2	3	4	5	6	7	8	9	10	11	12	
1	0	0	20	0	0	0	40	0	0	0	20	0	80
2	10	0	0	50	0	0	0	0	20	0	0	5	85

Note: i is the product type, k is the planning period

Table 9.6. Available capacity $c_j(k)$

j	\multicolumn{12}{c}{k}
	1	2	3	4	5	6	7	8	9	10	11	12
1	30	30	30	30	30	30	30	30	30	30	30	30
2	18	28	18	18	18	18	18	18	18	18	18	18
3	34	44	34	34	34	34	34	34	34	34	34	34
4	24	34	24	44	19	24	54	20	24	24	24	24
5	26	36	26	46	26	26	26	26	26	26	26	26
6	60	70	60	60	60	60	30	30	60	60	60	60
7	19	29	19	39	19	59	16	99	19	19	49	19
8	36	46	36	56	36	16	36	26	36	36	36	36
9	20	30	20	40	20	20	20	10	20	20	20	0
10	18	28	18	30	18	18	18	14	18	28	18	18

Note: j is the assembling stage, k is the planning period

Table 9.7. Production quantity $p_i(k)$, lot-size s_i and total production quantity

i	\multicolumn{12}{c}{k}	s_i	Total Production Quantity
	1	2	3	4	5	6	7	8	9	10	11	12		
1	10	0	10	0	15	5	5	5	10	15	5	0	5	80
2	10	10	10	20	0	10	0	10	10	0	0	10	10	90

Note: i is the product type, k is the planning period

$$\sum_{k=1}^{12} p_1(k) - \sum_{k=1}^{12} d_1(k) < s_1 \qquad (9.7)$$

and,

$$\sum_{k=1}^{12} p_2(k) - \sum_{k=1}^{12} d_2(k) < s_2 \qquad (9.8)$$

To further illustrate the effectiveness of the scheme, taking an example of Process 3 in Period 4 and the production schedule in the order indicated in Table 9.7, it can be found that there exists a shortage process that is

$$c_j(t)|_{j=3,t=4} < \sum_{i=1}^{2} w_{ij} \cdot d_i(t)|_{j=3,t=4} \qquad (9.9)$$

If the production is arranged according to the ETPSP as indicated in Table 9.5, the capacity shortage can be overcome due to the condition:

$$\sum_{i=1}^{2} w_{ij} \cdot p_i(t)|_{j=3,t=4} < c_j(t)|_{j=3,t=4} \qquad (9.10)$$

It is clearly demonstrated that the ETPSP is an effective means of solving the process capacity shortage problem by early or tardy production. Furthermore, Table 9.5 and Table 9.7 also indicate that the GA approach to ETPSP not only satisfies the customer requirement, but also minimizes the total early and tardy penalties.

The qualities of available capacity, order capacity requirement and balancing capacity requirement are shown in Fig. 9.4. The minimum earliness/tardiness penalty value for each iteration are shown in Fig. 9.6. Each trial represents a set of ETPSP for the manufacturing system. Fig. 9.6 shows the optimal individual in each generation. In this figure, the best individual of each generation is steadily converging to a near- optimal solution with the process of generations.

9.4.2 Case 2: Multi-product ETPSP

To further to demonstrate the GA approach, a reasonable large scale ETPSP problem is considered for further discussion.

Consider a manufacturing system consisting of 5 processes or assembling stages in a 6- period and producing 4-product for marketing. In the ETPSP formulation, three kinds of productions must proceed, i.e. two-product production, three-product production and four-product production. According to different products, the required production data are listed as follows:

9.4 GA Application for ETPSP 271

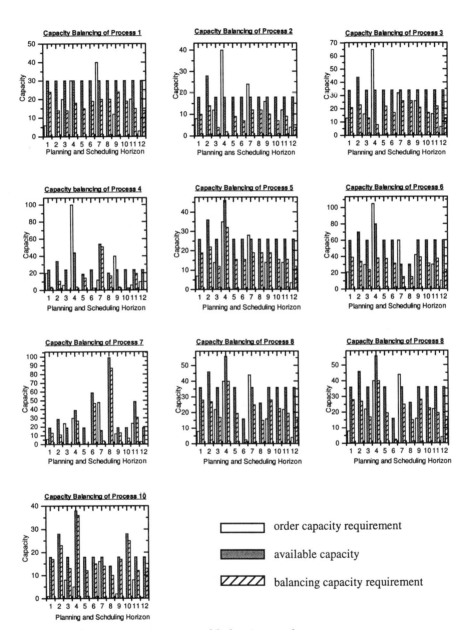

Fig. 9.5. Capacity requirement and balancing results

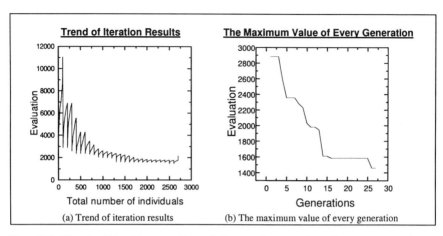

Fig. 9.6. Trend of iteration results and the maximum value of every generation

- Number of products N, respectively $N = 2$, $N = 3$ and $N = 4$;
- Number of processes $M = 5$;
- Length of a planning horizon $T = 6$;
- Earliness and tardiness penalties α_i and β_i, Lot-size $s_i (i = 1, 2, 3, 4)$, see Table 9.8;
- Order quantity $d_i(k)$ ($i = 1, 2, 3, 4$; $k = 1, 2, \cdots, 6$), see Table 9.9;
- Capacity requirement w_{ij} ($i = 1, 2, 3, 4$; $j = 1, 2, \cdots, 5$), see Table 9.10;
- Available capacity $c_j(k)$ ($j = 1, 2, \cdots, 5$; $k = 1, 2, \cdots, 6$), see Table 9.11.

Table 9.8. Earliness and tardiness penalties α_i and β_i

	Earliness Penalties α_i	Tardiness Penalties β_i	Lot-size s_i
Product 1	$\alpha_i = 5$	$\beta_i = 5$	$s_i = 5$
Product 2	$\alpha_i = 10$	$\beta_i = 20$	$s_i = 10$
Product 3	$\alpha_i = 5$	$\beta_i = 10$	$s_i = 5$
Product 4	$\alpha_i = 5$	$\beta_i = 15$	$s_i = 5$

Table 9.9. Order quantity $d_i(k)$

	Period 1	Period 2	Period 3	Period 4	Period 5	Period 6
Product 1	0	0	20	0	0	30
Product 2	10	0	0	50	0	0
Product 3	20	0	20	0	0	5
Product 4	0	0	10	40	0	10

For each of the production runs, 10 experiments were performed using different genetic parameters including population size, crossover rate and

9.4 GA Application for ETPSP

Table 9.10. Capacity requirement w_{ij}

	Process 1	Process 2	Process 3	Process 4	Process 5
Product 1	1.0	0.6	0.8	0.3	0.7
Product 2	0.6	0.8	1.3	2.0	0.7
Product 3	0.1	0.2	0.2	0.3	0.1
Product 4	0.3	0.2	0.1	0.4	0.2

Table 9.11. Available capacity $c_j(k)$

	Period 1	Period 2	Period 3	Period 4	Period 5	Period 6
Process 1	30	30	30	30	30	30
Process 2	18	28	18	18	18	18
Process 3	34	44	34	34	34	34
Process 4	24	34	24	44	19	24
Process 5	26	36	26	46	26	26

mutation rate.

Table 9.12 shows some average results for a population size of 30, while the crossover rate is 0.9 and mutation rate is 0.01. For the case where the population size is 100, the crossover rate is 0.6 and the mutation rate is 0.001. The obtained average results are stated in Table 9.13.

Table 9.12. 10 experiments with parameters 30, 0.9 and 0.01 by the GA approach

number of products N	average near-optimal value	average maximum generation	average running time	population size	crossover rate	mutation rate
N=2	715	14	94	30	0.9	0.01
N=3	3902.5	15	176	30	0.9	0.01
N=4	4750	18	235	30	0.9	0.01

Table 9.13. 10 experiments with parameters 100, 0.6 and 0.001 by the GA approach

number of products N	average near-optimal value	average maximum generation	average running time	population size	crossover rate	mutation rate
N=2	645	9	590	100	0.6	0.001
N=3	3450	12	678	100	0.6	0.001
N=4	4310.5	14	867	100	0.6	0.001

From Tables 9.12 and 9.13, it can be seen that when population size is 100, the average near-optimal value of 10 experiments is reached by the proposed GA approach. This is far better than the case where the population size is set to 30. Moreover, the average running time increases rapidly along with

274   9. Genetic Algorithms in Production Planning and Scheduling Problems

the increase of population size.

Taking Process 1 as an example, Fig. 9.7 shows the available capacity, capacity requirement and capacity balance results. From Fig. 9.7-(a), it can be found that in the whole production planning and scheduling horizon there are capacity shortages in Period 4 and Period 6, while capacity surpluses appear in other periods. This indicates that when the production is arranged according to the order, the manufacturer cannot satisfy the given order requirements due to capacity shortages in Period 4 and Period 6. Using the proposed GA approach, the ETPSP provides a means of obtaining the capacity balancing as shown in Fig. 9.7-(b). From this result, we see that there is no longer any capacity shortage in the entire planning and scheduling horizon.

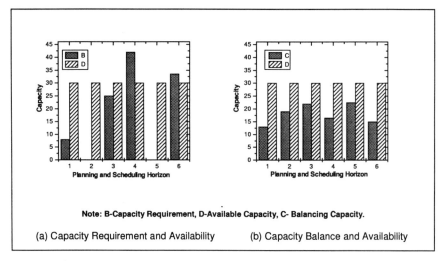

Fig. 9.7. Available capacity, capacity requirement and capacity balance of Process 1

The results obtained show that the GA approach to ETPSP not only satisfies the customers' requirement and capacity restraints, but also offers a near-minimum cost in terms of total early and tardy penalties. It can be demonstrated that the ETPSP is an effective means of solving the process capacity shortage problem brought about by early or tardy production.

In Fig. 9.8, the optimal individual is generated in each generation. Each trial represents a set of ETPSP for the manufacturing system. The best individual of each generation is steadily converging to a near-optimal solution with the process of generations.

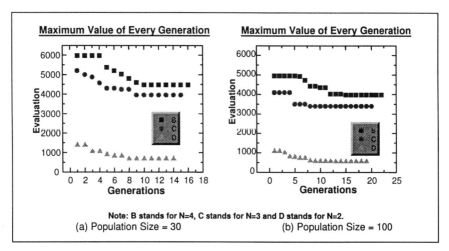

Fig. 9.8. Maximum of every generation

9.4.3 Case 3: MOGA Approach

In order to be able to support various and changing market requirements, the manufacturing systems should, in practice, be able to handle the involved processes and product-types with different lot-sizes. There are a number of design criteria, which consist of single and parallel multi-machine, multi-type product, due date, variable multi-process capacity instead of constant, as well as lot-sizes consideration. As ETPSP is essentially a large-scale MOGA problem, this could cause a short fall in the existing ETPSP methods when these criteria are not focused in a comprehensive manner.

Advanced MOGA Model of ETPSP. Objective Functions: To formulate ETPSP in the formation of MOGA, the objectives of concern are stated as follows:

1. Number of unbalancing processes,

$$f_1 = |P'|, \qquad (9.11)$$

and

$$P' = \left\{ j | j \in P_{key-process}, \ \sum_{i=1}^{N} w_{ij} p_i(k) - c_j(k) > 0, \ k = 1, 2, \cdots, T \right\} \qquad (9.12)$$

where $P_{key-process}$ is the set of the key-processes.

2. Cost of early production penalties,

$$f_2 = \sum_{i=1}^{N} \sum_{k=1}^{T} \alpha_i \left[l_i + \sum_{t=1}^{k} p_i(t) - \sum_{t=1}^{k} d_i(t) \right]^+ \qquad (9.13)$$

where $(x)^+ = \max\{0, x\}$.

3. Cost of tardy production penalties,

$$f_3 = \sum_{i=1}^{N}\sum_{k=1}^{T}\beta_i\left[\sum_{t=1}^{k}d_i(t) - \sum_{t=1}^{k}p_i(t) - l_i\right]^+ \quad (9.14)$$

where $(x)^+ = \max{0, x}$.

All the above objective functions must be minimized in order to achieve a satisfactory ETPSP based on MOGA.

Constraint Functions: Considering the process capacity balancing and production quality rationality, there are two groups of constraint functions stated as follows:

1. Process capacity constraint functions

$$\sum_{i=1}^{N} w_{ij}p_i(k) \leq c_j(k) \quad (9.15)$$

where $k = 1, 2, \cdots, T$, $j \in P - P_{key-process}$ and P is the set of all processes.

If we note,

$$s_{key-process} = |P_{key-process}| \quad (9.16)$$

then,

$$|P - P_{key-process}| = |P| - |P_{key-process}|$$
$$= M - s_{key-process} \quad (9.17)$$

From Eqn. 9.2, it can be seen that there are $(M - s_{key-process}) \times T$ constraint functions; and

2. Production quality constraint functions

$$0 \leq p_i(k), \quad p_i(k) \in S_i$$
$$S_i = \{r \cdot s_i, \ r = 1, 2, \cdots, \} \quad (9.18)$$

where $i = 1, 2, \cdots, N$ and $k = 1, 2, \cdots, T$.

Eqn. 9.5 indicates that each production quality must be positive. Meanwhile, it can be deduced that there are $N \times T$ constraint functions.

Preferential Ranking: During the process of optimization, the MOGA functions may not be minimized simultaneously. A Parato-based ranking technique is used to quantify the available chromosomes. For example, consider two individual chromosomes I_1 and I_2 with 3 objective values f_1^1, f_1^2, f_1^3 and f_2^1, f_2^2, f_2^3 respectively, I_1 is preferable to I_2 if and only if, (20)

$$f_2^1 = f_2^1 = 0 \tag{9.19}$$

$$\forall i = 2, 3; \quad f_1^i \le f_1^i \tag{9.20}$$

and

$$\exists j, \ j = 2, 3; \quad f_1^j < f_2^j \tag{9.21}$$

This ranking scheme provides an extra dimension in the strategy of optimization.

MOGA Approach to Multi-Process and Multi-product. A same 6–period MPSP problem handling for a 5–process, 4–product manufacturing system is considered here. The earliness and tardiness penalties, the order quantity, lot-sizes, the capacity requirements and available capacity of each process remain unchanged. By the use of a selection of key-processes, the most successful of these are obtained and listed Table 9.14.

Table 9.14. Key-processes

	Period 1	Period 2	Period 3	Period 4	Period 5	Period 6
Key-process	2	2 and 4	2	5	2	2

Some typical results are selected in Table 9.15. From this table, it can be clearly demonstrated that the MOGA approach is capable of making an effective ETPSP providing a multi-process capacity balance and producing multi-products for such a manufacturing system. The obtained MPSP not only minimizes the cost of early and tardy production penalties but also satisfies the capacity constraints.

Table 9.15. Performance of different objectives

	Objective f_1	Objective f_2	Objective f_3
N=2	0	605	475
N=3	0	1550	1370
N=4	0	2110	2200

Trade Off and Trends of Objectives. An important and extra feature of this MOGA approach is the trade off between the earliness/tardiness cost and the performance of the key-process balancing based on a minimum of objectives f_1 and f_2. It should be noted that this is only possible when the condition of objective $f_3 = 0$.

Without loss of generality, the two-product ($N = 2$) case is taken as an example. The minimum of the cost of early production against the minimum of the cost of tardy production is identified and shown in Fig. 9.9. It should also be noted the objective f_1 decreases and finally converges to zero (Fig. 9.10-(a)), while the objectives f_2 and f_3 converge to their near-optimal values as the generation increases (Fig. 9.10-(b) and Fig. 9.10-(c)). On the basis of the results obtained as shown in Figure 10, it can be concluded that

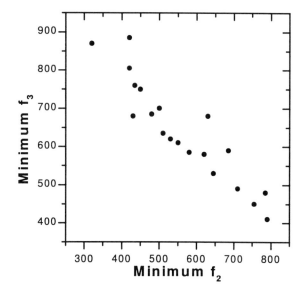

Fig. 9.9. Minimum of f_2 against minimum of f_3 in the final population

Different scale for ETPSP. To show the essence of the designed MOGA approach further, some simulation results have been achieved based on different scales of ETPSP according to different numbers of type-product. The obtained results are shown in Table 9.16.

Table 9.16. Achievable performance of different scale for ETPSP

	f_1	f_2	f_3	number of key-processes	number of processes	number of generations
N=2	0	605	475	26	69	30
N=3	0	1550	1370	26	69	30
N=4	0	2110	2200	28	69	32
N=5	0	2325	2515	29	69	40
N=10	0	5450	5675	34	69	80
N=20	0	13245	14220	38	69	120

Note: The length of a MPSP horizon $T = 24$, the number of processes $M = 69$, the ratio of crossover is 0.9, the ratio of mutation is 0.05, the population size=100.

It should be noted from Table 9.16, that the 5th column shows the numbers of key-processes among 69 processes during a 24-period horizon, and

9.4 GA Application for ETPSP   279

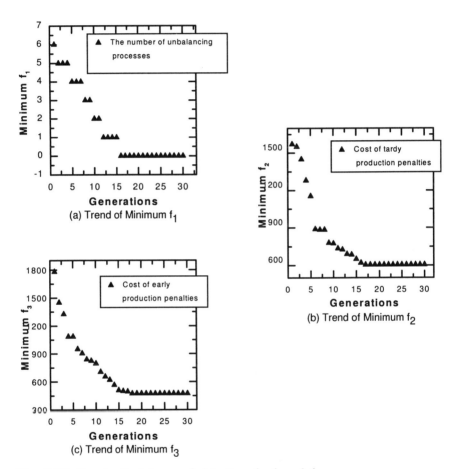

Fig. 9.10. Trends of minimum of objectives f_1, f_2 and f_3

280   9. Genetic Algorithms in Production Planning and Scheduling Problems

the last column shows the maximum generations when the iteration arrives at the optimal solutions. These results can only be obtained by the use of a MOGA approach due to its capacity to deal with the multi-type production along the MPSP horizon that has a multi-process capacity balancing ability. This provides a considerable and practical solution to practical ETPSP formulation.

Comparison. To illustrate the effectiveness of MOGA as compared with other techniques such as the key-process method (KPM), shrinking-feasible-field relaxation method (SFFRM) and the simple GA (SGA) approach, Table 9.17 shows the related functions for comparison. It is not difficult to see that both MOGA and Simple GA can achieve similar performance although MOGA's ability to handle the multiobjective functions clearly shows its advantage over the simple GA.

Table 9.17. Comparison among GA approach and other methods

Methods	*Aspects*				
	objective functions	lot-size consideration	capacity balancing	process number	product type
MOGA	Nonlinear, Multiple	Yes	Yes	Any	$N \geq 2$
SGA	Nonlinear, Single	Yes	Yes	Any	$N \geq 2$
KPM	Linear, Single	No	Yes	≤ 100	$N \leq 2$
SFFRM	Linear, Single	No	Yes	≤ 100	$N \leq 2$

9.5 Concluding Remarks

The MOGA approach to ETPSP formulation clearly demonstrates an effective means of tackling the multi-product environment. It provides solutions to large-scale multi-type production, with the considerations including the lot-size, and multi-process capacity balance for manufacturing. Since the unrealistic assumptions on the objectives such as linearity, convexity and differentiability, realistic scaling, are not required by MOGA for ETPSP, the results obtained are often genuine and considered as the optimal or at least near-optimal solutions.

In this way, manufacturers can respond to the changing market requirements in a timely manner and fulfill the need of customers. This method is a noted improvement on any existing techniques, and also in practice, provides a new trend for integrating the MRP-II and JIT processes.

10. Genetic Algorithms in Communication Systems

In communication systems, the common optimizing problems of concern are the capacity and delay constraints, routing assignment, topology and cost etc. These issues involve nonlinear and discrete functions for optimization which may not yield the adequate solutions easily using the classical gradient type of searching tools. The problem can be more severe when multiobjective functions are encountered.

This chapter aims to provide typical solutions to address the design dilemma by the use of GA as an optimizing and searching method. In this chapter, each of the design case studies is brought out the uniqueness of the system characteristics that can only be tackled by GA in a comprehensive manner is shown. In the first design study, GA is applied for the solving of a highly constrained optimization problem in an asynchronous transfer mode (ATM) network. The results obtained are compared directly with those obtained using an heuristic approach.

A similar approach is also applied to the design of mesh networks as demonstrated in the second design. In this particular study, the design foundation is developed upon the concept of reliability, in which case the failure of the nodes does not bring down the entire network. The employment of GA here is not only to obtain an optimized network topology but to solve problems in routing, the continuous and discrete capacity, and cost.

The final communication system that is of interest is the design of the Wireless Local Area Network (WLAN). In this design, HGA is applied to the minimization of the path losses between the terminals and base-stations. Because of HGA's capability for solving the skewed multiobjective functions and constraints, it can be found that a minimum number of base-stations is also identified.

10.1 Virtual Path Design in ATM

In broadband ATM networks, the cell based switching capacity is often built on top of the digital cross connect system (DCS) networks. The DCS networks provide the backbone for connecting the ATM switches and network reconfiguration. Hence the optimization problem is to configure the topology and to assign the capacity within the given facilities provided by the DCS network.

Consider the original (backbone) network shown in Fig. 10.2, the concept of an express pipe, which directly connects two ATM switches via the DCS network without intermediate ATM switching having to be introduced. For example, an express pipe can be established between A and C via P–T–S–R. Thus, different embedded topologies may be derived from the backbone with the establishment of different express pipes. Further configurations can then be made, in order to accommodate various traffic fluctuation due to the flexibility provided by DCS. The express pipes also decrease the number of intermediate switchings at the expense of reducing the multiplexing gains, as traffic is segregated onto different paths.

10.1.1 Problem Formulation

Having realized the DCS or backbone network, our objective is to obtain the topology of the embedded ATM network. The network optimization problem is formulated as a minimization of the congestion based on the average packet delay. The variables are the topology of the embedded network, the routing and the capacity assignment of different links in the embedded ATM networks.

The problem of designing the topology of an ATM network embedded in a DCS network [72] can be summarized in Table 10.1. As derived in [72], the average packet delay, excluding propagation delay, is contributed by two components: the switch's buffer overflow and the total link delay. The switch's buffer overflow probability is approximated by the trunk queueing delay, modeled as a simple M/M/1 queue. For the total link delay, [72] worked on the assumption that the ATM network consists of independent M/M/1 queues. Propagation delays are not included because they "depend on the geographical distribution of user sites, and are only marginally affected by the network topology layout" [72], and have little effect on the optimization. Hence, the average packet delay (in seconds) over the entire network excluding propagation delay is given by

$$T = \frac{1}{\lambda} \left[\sum_{m=1}^{\bar{M}} \frac{\bar{f}_m}{\bar{C}_m - \bar{f}_m} + \sum_{n=1}^{S} \frac{\mu \bar{f}_n}{\bar{K}_n - \mu \bar{f}_n} \right] \quad (10.1)$$

where $1/\mu$ is the packet length (in bits); $\bar{M}$ is the number of embedded links; $\bar{C}_m$ and $\bar{f}_m$ are the capacity and the aggregate flow on embedded link m, both in bits/second, respectively; $\lambda = \mu \sum_{k=1}^{Q} \lambda_k/\mu$ is the average offered flow of commodity k where Q is the number of commodities; S is the number of ATM switches; $\bar{K}_n$ is the throughput capacity of switch n in packets/second; and $\bar{f}_n$ is the aggregate flow through switch n in bits/second.

Table 10.1. Problem formulation

Given	Topology of backbone network Trunk capacity of the backbone network Switch Capacity of the backbone network Traffic flow requirement R
Minimize	Average packet delay
Variables	Topology of embedded network Routing on the embedded network Embedded link capacities
Subject to	Trunk capacity constraints Switch capacity constraints Satisfying R

10.1.2 Average packet delay

The average packet delay (in seconds) excluding propagation delay is given by

$$T = \frac{1}{\lambda} \left[\sum_{m=1}^{\bar{M}} \frac{f_m}{\bar{C}_m - f_m} + \sum_{n=1}^{S} \frac{\mu \bar{f}_n}{\bar{K}_n - \mu \bar{f}_n} \right] \tag{10.2}$$

where $1/\mu$ is the average packet length (in bits); $\bar{M}$ is the number of embedded links; $\bar{C}_m$ and $\bar{f}_m$ are the capacity and the aggregate flows on link m, both in bits/second, respectively; Q is the number of commodities; $\lambda = \mu \sum_{k=1}^{Q} r_k$ and $r_k = \lambda_k/\mu$ is the average offered flow of commodity k; S is the number of ATM switches; $\bar{K}_n$ is the throughput capacity of switch n in packets/second; and $\bar{f}_n$ is the aggregate flow through switch n in bits/second.

The capacity $\bar{C}_m$ of an embedded link is usually a discrete variable, i.e. multiples of 150 Mbits/s. This imposes the difficulty in the gradient optimization technique which assumes the continuity of the variables.

10.1.3 Constraints

In the virtual path design, there are a number of constraints to be considered.

284   10. Genetic Algorithms in Communication Systems

Trunk Capacity Constraint. The trunk capacity must not be exceeded by the aggregate capacity of all the second-order arcs that use the trunk.

$$\sum_{u=1}^{\bar{M}} \bar{C}_u p_{ui} \leq C_i \quad \forall i = 1, 2, \ldots, n \tag{10.3}$$

where $\bar{M}$ is the number of second-order arcs; p_{ui} equals "1" if the second-order arc $\bar{C}_u$ use the trunk i and p_{ui} equals "0" if otherwise.

Switch Capacity Constraint. The switch capacity should not be exceeded.

Flow Requirement. The aggregate flow on each second-order arc must not exceed its capacity.

$$\bar{f}_u \leq \bar{C}_u \quad \forall i = 1, 2, \ldots, \bar{M} \tag{10.4}$$

10.1.4 Combination Approach

Fig. 10.1 shows the block diagram for the virtual path design in the ATM net using the proposed approach. To implement the system, a combination of user-designed software and GENOCOP is adopted.

Routing Cycle. For a given backbone topology, the routing problem will be solved by the GA. The routing scheme is considered as the GA chromosome, P, represented in a hierarchical manner:

$$P = \{p_1, p_2, \ldots, p_{n(n-1)/2}\} \tag{10.5}$$

where p_k is the gene that presents the routing path from node i to j; $i = 1, \ldots, (n-1)$; $j = (i+1), \ldots, n$; $k = g(i,j) = (i-1) \cdot n - i \cdot (i+1)/2 + j$.
p_k is formulated as follow:

$$p_k = \begin{cases} [x_0, x_1, x_2, \ldots, x_e] \\ [b_1, b_2, \ldots, b_e] \end{cases} \tag{10.6}$$

where b_m is a bit to determine whether the link connecting x_{m-1} and x_m is an expressed pipe or not. If b_m equals to 1, then the link is an expressed pipe, or vice versa.

With a particular routing scheme P, we can generate the embedded topology and the flow of the virtual paths according to the traffic requirement R. A number of equalities can be formulated based on the trunk capacities and switch capacities given. If the constraints are violated, a penalty value is given in the objective value. Otherwise, the fitness of the routing scheme can

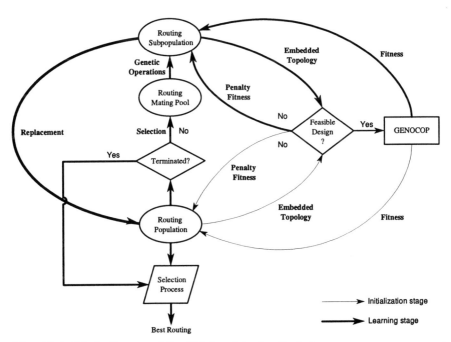

Fig. 10.1. Block diagram of the GA for virtual path design

be evaluated using the optimal solution obtained from the GENOCOP[149] (see Fig. 10.1).

The crossover operation is explained as follows:
With the parental chromsomes P_1 and P_2

$$P_1 = \{p_1, p_2, \ldots, p_{n(n-1)/2}\}$$
$$P_2 = \{p'_1, p'_2, \ldots, p'_{n(n-1)/2}\}$$

The offspring generated is

$$P = \{s(p_1, p'_1), s(p_2, p'_2), \ldots, s(p_{n(n-1)/2}, p'_{n(n-1)/2})\}$$

where

$$s(p_i, p_j) = \begin{cases} p_i & \text{if } p_i \text{ has shorter distance than } p_j \\ p_j & \text{otherwise} \end{cases}$$

For mutation, the routing path p_k will be randomly re-routed if the probability test is passed. The operation rate is set at 0.05.

Capacity Assignment Cycle. The use of GENOCOP is due to its ability to handle the linear constraints (equations and inequalities) in this problem. Since the capacity $\bar{C}_m$ of an embedded link is usually a discrete variable, the GENOCOP is modified so as to use it in the integer domain.

10.1.5 Implementation

The template for the ATM problem is the same as shown in Table. 10.17. Since the GENOCOP is used so that the constraints can be easily handled, some modifications are listed as follows:

10.1.6 Results

The same example as in [72] is used to evaluate the proposed approach. The backbone network is depicted in Fig. 10.2. Each trunk and ATM switch have the capacity of 16×150 Mbit/s and 32 packet/s, respectively. GA applied here for the traffic requirement as tabulated in Table 10.2.

The processing time used was about 32.09s on a Pentium-100. The optimal embedded network obtained is depicted in Fig. 10.3. The average packet delay of this embedded network is only 1.371s, which is less than the optimal result 1.373s obtained in [72], taking 30 minutes in SUN 3/280.

Class Chromosome
    DEFAULT_SIZE = number of ATM nodes
    DEFAULT_CTL = 1

Class GeneticAlgorithm
    void GeneticAlgorithm::Evaluate(Population Pop)
    {
        int DEF_SIZE = 20 ;   // Population size in GENOCOP GA
        G_GA GENOCOP_GA ;   // G_GA Class in GENOCOP
        G_Population GENOCOP_P ;   // G_Population, Class in GENOCOP

        Chromosome chrom ;   // Chromosome in user-design class
        Population TempPop(Pop.getSize()) ;

        for (i =0 ; i < Pop.getSize(); i++)
          {
            // Conversion the user-design structure to GENOCOP chromosome structure
            G_Population(DEF_SIZE) ;
            // Define A Population in GENOCOP
            GENOCOP_GA() ;
            // Using GENOCOP to fine the optimal routing scheme.
            G_CHROM = G_Population.Best() ;
            Conv_GENOCOP_2_chrom(G_CHROM,chrom) ;
            // Conversion the GENOCOP chromosome structure to user-design structure
            TempPop.add(chrom) ;
            // Add the optimal solution obtained from GENOCOP to temp population
          }
        Pop.copy(TempPop) ;
        Update_Fitness(Pop) ;
    }

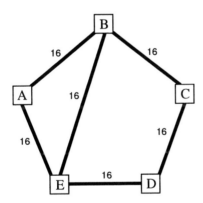

Fig. 10.2. Backbone topology

Table 10.2. Original traffic matrix

	A	B	C	D	E
A	–	4.0	4.0	0.5	6.5
B	4.0	–	4.0	0.0	4.0
C	4.0	4.0	–	4.0	0.5
D	0.5	0.0	4.0	–	4.0
E	6.5	4.0	0.5	4.0	–

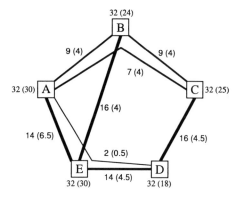

Fig. 10.3. Best network for original traffic change

Supposing that there is a demand of traffic between nodes B and D, assumed to be 3.0, another optimal second order network would exist. The re-configuration was completed after 42.28s in the same machine. The final embedded network, with the average packet delay 1.712s, is depicted in Fig. 10.4.

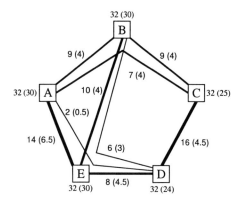

Fig. 10.4. Best network obtained after traffic change

10.2 Mesh Communication Network Design

In general, the designs for mesh communication networks are complex, multiconstraint optimization problems. The designs are used to find optimal assignments of link capacity, routing and topological connections such that the cost is minimized and yet satisfies the traffic requirements and the maximum permissible delay. In fact, the problem is *NP*-complete and for a practical problem with a modest number of nodes, only approximate solutions can be obtained through heuristic algorithms.

Currently, the heuristic methods used in the design of mesh networks include branch-exchange, cut-saturation, and the more efficient MENTOR algorithms [122]. In the design of mesh networks, the component subproblems of routing and capacity assignments are also difficult problems. In particular, the relation between the cost and capacity is often assumed to be linear. The more likely concave cost-capacity function is often modelled by a line tangent to the function. For the more realistic discrete cost-capacity function, many of the current design methods can only provide an approximate solution with a bound to the optimum. For a method which guarantees an optimal solution with discrete cost-capacity function, this will require a very long searching time to provide a realistic size of network (Chapter 7 of [122]).

Previous works in this application are few and restrictive in their formulation. The problem of the optimal communication spanning tree using GA is shown in [164], and a solution for local area networks in [55]. The design of packet switched networks using GA is attempted in [168]. However, only the network topology has been optimized in this approach, and no attempt has been proposed for the optimization on routing and capacity.

In this section, the C++ library was applied to find a total solution for the networking problem. Not only was the network topology optimized, but the optimization of capacity and routing were also included in our task. We used a mesh packet switched communication network that is based on the formulation stated in [71] for such a design.

Because of the intrinsic characteristics of GA [139], which has a unique capability for handling discrete events, multiobjective functions, constraints etc, the conflicting inter-relationship between the network topology, routing and capacity (cost) presents no difficulty for GA. Considering that the network design is largely governed by the specification that is formulated in Table 10.3, the whole complexity of the problem is broken down and driven only by the constraints and the requirements of cost. This can be simplified into three sets of design variables as indicated in Table 10.3 which corresponds to the number of optimization levels, as depicted in Fig. 10.5.

In this way, each optimization level has the main core and a GA cycle, with similar architecture. This similarity can largely reduce the complexity of the system design. The advantages of using this approach are not only its elegance and simplicity, but also its ability to handle continuous and discrete link capacities, linear or discrete cost structures, additional constraints and

Table 10.3. Formulation of a topological design problem

Given	Node locations Traffic requirements matrix R between node pairs
Minimize	Total connection cost
Over design variables	Topology Routing Channel capacities
Subject to	Link capacity constraints Average delay constraints Reliability constraint

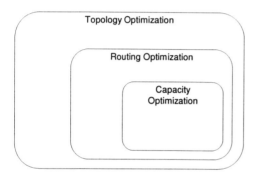

Fig. 10.5. 3-level optimization process

different constraint models.

10.2.1 Design of Mesh Communication Networks

In order to formulate the GA method for mesh communication networks, the essential network architecture and its associated design parameters for investigation should be clearly defined. This section also outlines the features of concern.

Network Modelling. In a packet-switched network, variable size packets are stored-and-forwarded by the switches or nodes. Traffic arrives at the nodes in a stochastic process and queues of packets are stored at the buffers of the switches. Data are lost if the buffers overflow. Assuming that congestion rarely occurs or data are seldom lost in a well designed network, the packet switched network can be modelled as a network of queues with infinity buffers or a delay system.

The average delay in each of the link of the network can be approximated by the M/M/1 model based on the Kleinrock independence approximation (see [21]). The average delay is given by

$$T = \frac{1}{\gamma} \sum_{i=1}^{b} \frac{f_i}{C_i - f_i}, \qquad (10.7)$$

where γ is the total arrival rate into the network; b is the number of links; f_i, C_i are the assigned link flow and capacity of link i, respectively.

Let $d_i(C_i)$ be the cost of leasing C_i in link i, the total connection cost is simply

$$D = \sum_{i=1}^{b} d_i(C_i). \qquad (10.8)$$

It is interesting to note that the solution method using GA does not depend on the modelling of the delay or the cost structure. One can simply change the delay modelling technique or the cost structure for different networks or requirements.

Reliability Consideration. In order to improve the reliability of a network, the network must not contain any single link or node whose breakdown would lead to disconnection of the network. If a node is found such that its breakdown disconnects the network into two or more pieces, the links will need to be redesigned in order to improve reliability.

In our reliability requirement, the network is biconnected, i.e. there are 2 node-disjoint paths connecting every two nodes in the network. Such a reliability requirement will give rise to a mesh network since adjacent nodes are interconnected. The depth-first search [193] is adopted for testing the biconnectivity of the network. The time complexity of this algorithm is related to $n + b$ where n and b are the number of nodes and links, respectively.

If the reliability constraint of biconnectivity is relaxed, the minimum cost network will have the structure of a minimum spanning tree. Higher reliability requirement can be specified by increasing the connectivity of each node, e.g. k-connected, or limited by the number of hops between any two nodes. These reliability requirements can also be built into our model.

10.2.2 Network Optimization using GA

The overall optimisation strategy is largely governed by the 3-level optimization process indicated in Fig. 10.6. A detailed account for each level of optimization is given in the following subsections.

10. Genetic Algorithms in Communication Systems

Topology Optimization. In this optimization process, the topology of the network need not be specified. The set of links connecting the nodes is chosen in order to define the network topology.

Table 10.4. Formulation of a topology design problem

Given	Traffic requirements matrix R
Minimize	$D(A, C) = \sum_{i \in A} d_i(C_i)$
	where the set of arcs A specifies the topology
Subject to	$\mathbf{f} \leq \mathbf{C}$
	$T = \frac{1}{\gamma} \sum_{i=1}^{b} \frac{f_i}{C_i - f_i} \leq T_{max}$
	$\mathbf{f}$ is a multicommodity flow satisfying R
	Set A must correspond to a biconnected topology

The node name is mapped to node number (X) which is an integer between 1 and n. The topology is defined by the sets of the node and edges connecting the nodes. The most straightforward representation for topology is the so-called *adjacency matrix* representation. A $n \times n$ array of Boolean values is maintained, with $a[x][y]$ set to 1 if there is an edge connecting from node x to node y, and 0 if otherwise. It is assumed that the link is bi-directional, which means that a symmetric matrix is obtained. Hence, only $n(n-1)/2$ binary numbers are required to represent the topology.

For the sake of simplicity, the two-dimensional matrix is transformed into a one-dimension matrix A with the following function:

$$A[k] = a[i][j] \quad (10.9)$$

where $j > i$ and $k = g(i,j) = j \cdot X - (j \cdot (j+1)/2) + i - j - 1$

Hence, the topological chromosome can be formulated as a binary string with $n(n-1)/2$ elements so as to represent the one-dimensional topology matrix A in Eqn. 10.9. The crossover and mutation employ the conventional operation of one-point crossover and the bit mutation, respectively.

This optimization procedure, based on the design specification stated in Table 10.4, cannot be completed alone without the involvement of routing and capacity assignments. These become available at the end of their own respective GA cycles, which are described later. The final topology is reached when the optimal cost values are obtained. The general GA cycle of this level of operation is depicted in Fig. 10.6.

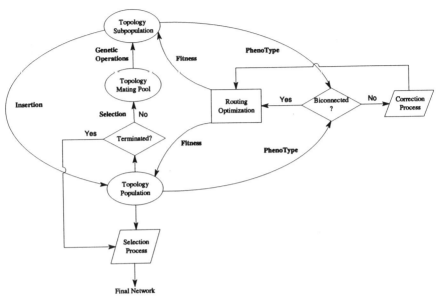

Fig. 10.6. Topology design optimization process

Routing Optimization. For a given topology, the capacity assignment and routing problem requires the simultaneous optimization of both the flow and link capacities as defined in Table 10.5.

In this problem, we consider the routing scheme as the GA chromosome. The routing chromosome is then represented by a path list:

$$chromosome: \qquad P = \{p_1, p_2, \ldots, p_{n(n-1)/2}\} \qquad (10.10)$$

where $p_k = Path(i,j) = [i, x_1, x_2, \ldots, x_e, j]$ is the gene that presents the routing path from node i to j; $i = 1, \ldots, (n-1)$; $j = (i+1), \ldots, n$; $k = g(i,j)$

With a particular routing scheme P, the flow of the link can be assigned according to the traffic requirement R. Hence, the fitness of the routing scheme can be evaluated using the optimal solution obtained from the capacity assignment optimization. The GA cycle of the routing optimization is shown in Fig. 10.7.

Crossover. With the parental chromsomes R_1 and R_2

Table 10.5. Formulation of a capacity assignment and routing problem

Given	Topology Traffic requirements matrix R
Minimize	$D(\mathbf{C}) = \sum_{i=1}^{b} d_i(C_i)$
Over design variables	$\mathbf{f}, \mathbf{C}$
Subject to	$\mathbf{f} \leq \mathbf{C}$
	$T(\mathbf{f}, \mathbf{C}) = \frac{1}{\gamma} \sum_{i=1}^{b} \frac{f_i}{C_i - f_i} \leq T_{max}$
	$\mathbf{f}$ is a multicommodity flow satisfying R

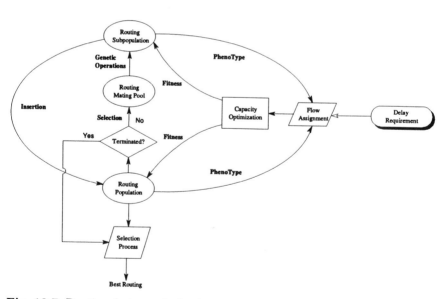

Fig. 10.7. Routing design optimization process

$$P_1 = \{p_1, p_2, \ldots, p_{n(n-1)/2}\}$$
$$P_2 = \{p'_1, p'_2, \ldots, p'_{n(n-1)/2}\}$$

The offspring generated is

$$P = \{s(p_1, p'_1), s(p_2, p'_2), \ldots, s(p_{n(n-1)/2}, p'_{n(n-1)/2})\}$$

where

$$s(p_i, p_j) = \begin{cases} p_i & \text{if } p_i \text{ has shorter distance than } p_j \\ p_j & \text{otherwise} \end{cases}$$

Mutation. The routing path p_k will be randomly re-routed if the probability test is passed. The operation rate is set at 0.05.

Capacity Assignment Optimization. For a complete topological network design, the capacity assignment determines the quality of the network. Although the methodology of structuring topology and routing schemes has been introduced, the design has yet to be completed without evaluating the cost of the network. Similar to the previous two GA cycles of optimization routines, the same approach applies.

The capacity assignment problem is to choose the capacity C_i for link i such that the network cost is minimal, subject to the maximum average delay constraint and the assigned flow requirements. The capacity assignment problem is defined in Table 10.6.

Table 10.6. Formulation of a capacity assignment problem

Given	Topology Traffic requirements matrix R Routing, or flow vector $\mathbf{f} = (f_1, f_2, \cdot, \cdot, f_b)$
Minimize	$D = \sum_{i=1}^{b} d_i(C_i)$
Over design variables	$\mathbf{C} = (C_1, C_2, \cdot, \cdot, C_b)$
Subject to	$\mathbf{f} \leq \mathbf{C}$ $T = \frac{1}{\gamma} \sum_{i=1}^{b} \frac{f_i}{C_i - f_i} \leq T_{max}$

In general, the smaller the maximum average delay, the higher will be the link capacities required. Hence, a Pareto optimal set exists between the average delay and the required capacity or connection cost.

In order to identify the Pareto optimal set, a multiobjective GA approach is adopted. The delay is not considered as a constraint, but as another objective function for minimization. The GA cycle is shown in Fig. 10.8.

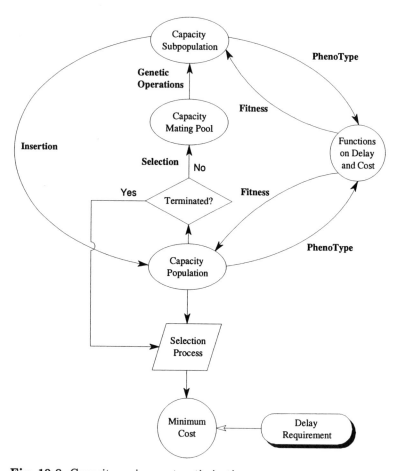

Fig. 10.8. Capacity assignment optimization process

The linkage chromosome structure is

$$C = \{C_1, C_2, \ldots, C_b\} \tag{10.11}$$

where b is the number of links.

For the continuous capacity assignment problem,

10.2 Mesh Communication Network Design

$$C_k \in \Re \geq 0 \quad k = 1, 2, \ldots, b \qquad (10.12)$$

For a discrete capacity assignment problem, each element C_k is an integer array of dimension m which is the number of capacity types.

$$C_k = \{C_k[1], C_k[2], \ldots, C_k[m]\} \qquad (10.13)$$

where $C_k[i] \in Z \geq 0$ is the number of capacity type i used in the link k. Combination of different capacity types is allowed on each link.

The genetic operations are one-point crossover and random mutation.

Example 1:
To demonstrate the effectiveness of this approach of topological design, a 5-node network is used. The topology and flow requirements are shown in Fig. 10.9 and Table 10.7, respectively. Continuous capacity assignment is assumed.

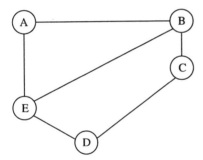

Fig. 10.9. Example

Table 10.7. Flow requirement and cost/cap of Example 1

Endpoints		Flow	Cost/Cap
A	E	3	700
A	B	1	1600
E	B	6	1500
E	D	2	1100
B	C	4	500
C	D	5	1300

The delay and the connection cost of those rank 1 chromosomes in the final generation are depicted in Fig. 10.10. A Pareto optimal set is clearly obtained by the GA approach.

298   10. Genetic Algorithms in Communication Systems

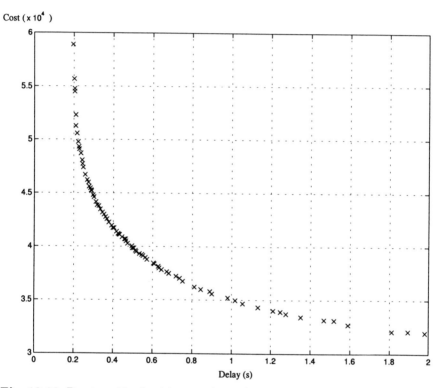

Fig. 10.10. Pareto optimal set in capacity assignment

10.2.3 Implementation

[Parameters and modification] To implement the topology optimization, new data members and functions are added to the *Chromosome Class*.
The additional data members and functions in chromosome Class are shown below:

```
data members
                // Total Capacity of the Topology
                double     TCapacity ;      protected
                // Total Cost of the Topology
                double     TCost ;          protected
                // Total Delay of the Topology
                double     TDelay ;         protected
                // Read and write access of the edge between node x and y.
Edge            virtual void Edge(int x, int y, bool ON_Off) ;
                virtual bool Edge(int x, int y) ;
                // Read and write access of the topology flow between node x and y
Flow            virtual void Flow(int x, int y, double flo) ;
                virtual double Flow(int x, int y) ;
                // Read and write access of the topology capacity between node x and y
Cap             virtual void Cap(int x, int y, double) ;
                virtual double Cap(int x, int y) ;
                // return the number of steps from src to dest, and the travelling path
Travel          virtual Travel(int src, int dest, int* Step, int& Path) ;
                // Read access of Capacity, Cost and Delay
getTCapacity    virtual double getTCapacity() const ;
getTCost        virtual double getTCost() const ;
getTDelay       virtual double getTDelay() const ;
```

The additional data members and functions in *Population Class* :

```
DReinsert  Population DReinsert(Population P, Population SubP) ;
           Reinsert the population, P, from the sub-population, SubP.
           Base on the Chromosome (Topology) TDelay and the Delay requirement.
CReinsert  Population CReinsert(Population P, Population SubP) ;
           Reinsert the population, P, from the sub-population, SubP.
           Base on the Chromosome, Topology, TCapacity and the Capacity requirement.
TReinsert  Population TReinsert(Population P, Population SubP) ;
           Reinsert the population, P, from the sub-population, SubP.
           Base on the Chroosome, Topology, TCost and the Cost requirement
```

The additional data members and functions in *GeneticAlgorithm Class* :

Network Topology Optimization Template. The Network optimization governed by the 3-level optimization process, the inter-level optimization process coding, Fig. 10.5, as shown in the following :

300   10. Genetic Algorithms in Communication Systems

DEvaluate virtual void DEvaluate(Population P) ;
    Calculate the Delay of the Chromosome (Topology) in the population.
CEvaluate virtual void CEvaluate(Population P) ;
    Calculate the Capacity of the Chromosome (Topology) in the population.
TEvaluate virtual void TEvaluate(Population P) ;
    Calculate the Cost of the Chromosome (Topology) in the population.

Table 10.8. A template for delay optimization

```
Chromosome GAs_Delay(Chromosome Parent)
{
double xRate = 0.01 ;                          // crossover Rate
double mRate = 0.02 ;                          // mutation Rate
int Population_Size = 20 ;                     // Population size
int Sub_Population_Size = 20 ;                 // Sub Population size

Population Pop(Population_Size, Parent)        // Set all the chromosomes in the
                                               // population with same topology
                                               // structure of the Parent
Population SubPop(Sub_Population_Size) ;

Genetic Algorithm Capacity(Pop,xRate,mRate) ;
Capacity.DEvaluate(Pop) ;

do
  {
       Capacity.DSelect_Parent(Pop,SubPop) ;
       Capacity.DRecombine(SubPop) ;           // keep the network topology
       Capacity.DMutate(SubPop) ;
       Capacity.DEvaluate(SubPop) ;
       Capacity.DReinsert(Pop,SubPop) ;
       Capacity.DGeneration() ;
  } While(Capacity.DTerminate = TRUE ) ;

  return Capacity.getbest()
} // End Delay Optimization
```

10.2.4 Results

Having now firmly established the design methodology for communication networks using GA, a design study is carried out for the design of a mesh packet switched communication network that is proposed between Hong Kong and China. This is a 10-node Chinese network which is shown in Fig. 10.11 with the forecast traffic requirement shown in Table 10.11. The cost structure is assumed to be proportional to the distance and consists of three different line-rates with unit cost per kilometer as given in Table 10.12. The distances between these 10 major Chinese cities in kilometers are shown in Table 10.13. A biconnected network is required to be designed with a maximum allowable packet delay of 0.1s.

This matrix of line-rates which, in general, imposes additional constraints on the conventional search methods can be used in favour of the GA approach. It can reduce the searching domain of C_k in Eqn. 10.13 by noting the following

10.2 Mesh Communication Network Design

Table 10.9. A template for routing optimization

```
Chromosome GAs_Routing(Chromosome Parent)
{
double xRate = 0.01 ;                        // crossover Rate
double mRate = 0.02 ;                        // mutation Rate
int Population_Size = 20 ;                   // Population size
int Sub_Population_Size = 20 ;               // Sub Population size

Population Pop(Population_Size, Parent) ;    // Set all the chromosome with
                                             // same topology and capacity
                                             // structure of the Parent
Population SubPop(Sub_Population_Size) ;

Genetic Algorithm Routing(Pop,xRate,mRate) ;
Routing.REvaluate(Pop) ;

do
  {
      Routing.RSelect_Parent(Pop,SubPop) ;
      Routing.RRecombine(SubPop) ;
      Routing.RMutate(SubPop) ;
      Routing.REvaluate(SubPop) ;
      Routing.RReinsert(Pop,SubPop) ;
      Routing.RGeneration() ;
  } While(Routing.RTerminate = TRUE ) ;

  return Routing.getbest()
} // End Routing Optimization
```

cost-capacity combinations from Table 10.12.

- $0 \leq C_k[2] \leq 2$;
- if $0 \leq C_k[2] \leq 1$ then $0 \leq C_k[1] \leq 3$;
- if $C_k[2] = 2$ then $C_k[1] = 0$.

Using the GA formulation in Fig. 10.5, the final topology and the corresponding routing scheme obtained are shown in Fig. 10.12 and Table 10.15. The capacity assigned for each link is also shown with the assigned link flows in parenthesis. The results of the final topology are listed in Table 10.16. For comparison, the method of branch exchange is applied. Starting with the complete graph, a link is dropped if the cost per bit is highest. This process is repeated until there is no improvement. The routing is based on the minimum distance route and the capacity is assigned with delay constraints fulfilled based on the heuristic approach. The final topology is depicted in Fig. 10.13. The results using branch exchange are summarized in Table 10.16 which is much inferior to the GA approach. A mentor approach is not applicable due to the requirement of 2-connectivity and the fact that the some links between nodes are not available from the original graph.

302   10. Genetic Algorithms in Communication Systems

Table 10.10. A template for typical GA application

```
// Get the optimal network for a given Network Topology
Network_Evaluate(Population Pop)
{
    Population TempPop(Pop.getSize()) ;
    for (i = 0 ; i<Pop.getSize(); i++)
        TempPop.add(GAs_Routing(GAs_Delay(Pop.member(i)))) ;
    Pop.copy(TempPop) ;
}   // End Network_Evaluate


void main()
{
double xRate = 0.01 ;                          // crossover Rate
double mRate = 0.02 ;                          // mutation Rate
int Population_Size = 20 ;                     // Population size
int Sub_Population_Size = 20 ;                 // Sub Population size

Population Pop(Population_Size) ;
Population SubPop(Sub_Population_Size) ;
Genetic Algorithm GAs(Pop,xRate,mRate) ;
Topology.TEvaluate(Pop) ;
do
  {
      Topology.TSelect_Parent(Pop,SubPop) ;
      Topology.TRecombine(SubPop) ;
      Topology.TMutate(SubPop) ;               // network topology mutation
      Network_Evaluate(SubPop) ;               // get the optimal network
                                               // of a given topology
      Topology.TEvaluate(SubPop) ;             // network cost evaluation
      Topology.TReinsert(Pop,SubPop) ;         // network topology reinsert
      Topology.TGeneration() ;
  } While(Topology.Terminate = TRUE ) ;
} // End Main
```

Table 10.11. Traffic requirement between the 10 major cities (in Mbps)

	B	S	G	H	W	C	X	K	Ha	T
B	0	20	20	20	20	10	10	2	5	20
S	20	0	20	20	20	5	5	2	1	20
G	20	20	0	20	10	5	5	5	1	5
H	20	20	20	0	10	5	2	2	1	5
W	20	20	10	10	0	5	5	0	1	5
C	10	5	5	5	5	0	5	2	0	2
X	10	5	5	2	5	5	0	0	0	2
K	2	2	5	2	0	2	0	0	0	0
Ha	5	1	1	1	1	0	0	0	0	5
T	20	20	5	5	5	2	2	0	5	0

Table 10.12. Different line-rates

Type	Mbps	unit cost / km
1	6	1
2	45	4
3	150	9

10.2 Mesh Communication Network Design   303

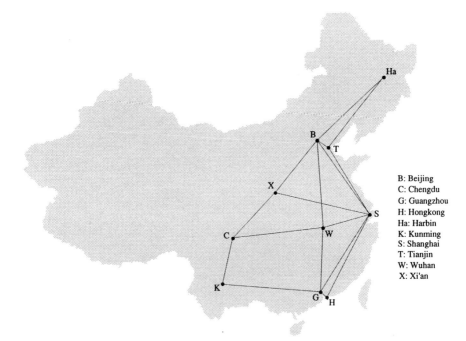

Fig. 10.11. Existing connections between 10 major cities in China

Table 10.13. Distance between the 10 major cities (in km)

	B	S	G	H	W	C	X	K	Ha	T
B	0	1200	2000	2100	1120	1600	960	2160	1120	160
S	1200	0	1280	1250	720	1680	1240	2000	2240	1120
G	2000	1280	0	240	840	1240	1360	1160	3120	2400
H	2100	1250	240	0	960	1480	1440	1400	3220	2370
W	1120	720	840	960	0	1000	680	1320	2190	1040
C	1600	1680	1240	1480	1000	0	640	680	2720	1600
X	960	1240	1360	1440	680	640	0	1240	2080	960
K	2160	2000	1160	1400	1320	680	1240	0	3280	2200
Ha	1120	2240	3120	3220	2190	2720	2080	3280	0	1150
T	160	1120	2400	2370	1040	1600	960	2200	1150	0

Table 10.14. Capacity and flow assignment between the 10 major cities using GA

Endpoints		Capacity / Mbps	Flow / Mbps
B	T	150	94
B	X	45	41
B	W	45	36
B	Ha	12	6
S	T	150	98
S	W	45	40
S	H	150	119
G	H	150	92
G	K	45	31
C	K	45	26
C	X	45	31
Ha	T	12	8

304   10. Genetic Algorithms in Communication Systems

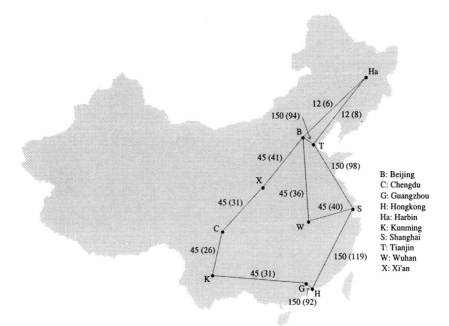

Fig. 10.12. Final topology between 10 major cities in China using GA

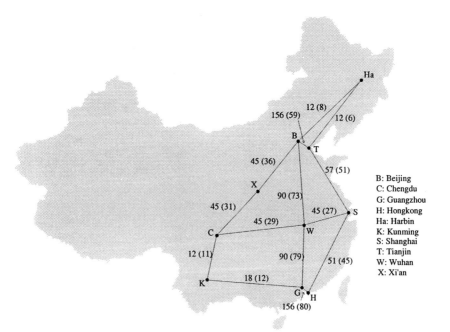

Fig. 10.13. Final topology between 10 major cities in China using branch exchange

Table 10.15. The routing scheme

Endpoints		Intermediate nodes				
B	S	B	T	S		
B	G	B	T	S	H	G
B	H	B	T	S	H	
B	W	B	W			
B	C	B	X	C		
B	X	B	X			
B	K	B	X	C	K	
B	Ha	B	Ha			
B	T	B	T			
S	G	S	H	G		
S	H	S	H			
S	W	S	W			
S	C	S	H	G	K	C
S	X	S	T	B	X	
S	K	S	H	G	K	
S	Ha	S	T	Ha		
S	T	S	T			
G	H	G	H			
G	W	G	H	S	W	
G	C	G	K	C		
G	X	G	K	C	X	
G	K	G	K			
G	Ha	G	H	S	T	Ha
G	T	G	H	S	T	
H	W	H	S	W		
H	C	H	G	K	C	
H	X	H	G	K	C	X
H	K	H	G	K		
H	Ha	H	S	T	Ha	
H	T	H	S	T		
W	C	W	B	X	C	
W	X	W	B	X		
W	K	W	S	H	G	K
W	Ha	W	B	Ha		
W	T	W	B	T		
C	X	C	X			
C	K	C	K			
C	Ha	C	X	B	Ha	
C	T	C	X	B	T	
X	K	X	C	K		
X	Ha	X	D	Ha		
X	T	X	B	T		
K	Ha	K	C	X	B	Ha
K	T	K	C	X	B	T
Ha	T	Ha	T			

Table 10.16. Summary of the final topology

	GA	Branch Exchange
Total Capacity	894Mbps	834Mbps
Total Cost	50590 units	55310 units
Delay	0.0644s	0.0999335

10.3 Wireless Local Area Network Design

In this section, a design of WLAN using HGA method is adopted. Based on the HGA formulation, it can be found that HGA is ideally suited for WLAN design in that it not only satisfies the optimization of the skewed multiobjective functions and constraints, but also a precise number of minimum required base-stations is identified. This added feature provides a design trade-off between cost and performance without requiring any extra effort.

10.3.1 Problem Formulation

Consider that the terminal locations of a WLAN are distributed over a designated region in a 3-D Euclidean space, it would be desirable to be able to determine the required number of base-stations as well as their precise locations so that the best quality of service of the network could be obtained. To achieve this objective, there are a number of technical problems to be overcome. Moreover, it is assumed that the capacity of the base-station is much larger than the traffic intensity of the allocated terminals and all base stations have equal power output.

Path Loss Model. The quality of the network is highly related to the path loss function of the terminals. This is generally governed by a mean path loss function in terms of distance with respect to the n-th power [175, 189].

$$S(d) \propto \left(\frac{d}{d_0}\right)^n \tag{10.14}$$

where S is the mean path loss, d_0 is a reference distance chosen as 1 metre; d is the distance between the terminal and the base-station, and n is the mean path loss exponent, indicating how rapidly the path loss is being dissipated as the distance increases. Indeed, n is a variable and subject to factors such as building type, layout, and the number of floors between base-station and terminal.

Hence, the absolute mean path loss, g_i, for a particular terminal i in decibels, can be computed as

$$g_i = S_0 + 10.0 \cdot n \log(d_i) \tag{10.15}$$

where d_i is the distance between the terminal i and the base-station; S_0 is due to free space propagation from the base-station to a 1m reference distance or $S_0 = 10 \cdot n_0 \log(4\pi \cdot 1m/\lambda)$ with $n_0 = 2$ and λ is the wavelength of the frequency in used; e.g. $S_0 = 37.55 dB$ at 1.8GHz.

To take into account any physical obstructions that lie directly between the base-station and the terminal, g_i can be modified as

$$\begin{aligned} g_i &= S_0 + 10.0 \cdot n_0 \log(d_i) + \sum_{w=1}^{M} N_w(i) L_w \\ &= 20.0 \log\left(\frac{4\pi d}{\lambda}\right) + \sum_{w=1}^{M} N_w(i) L_w \end{aligned} \qquad (10.16)$$

where $N_w(i)$ is the number of obstructing objects (for example walls) with type w separating the terminal i and the base-station; L_w is the penetration loss due to an obstructing object of type w and there are a total M types of objects.

Note that the free space exponent used in Eqn. (10.16) assumes that free space propagation applies for all distances [194].

For the multiple base-stations problem, a sub-problem of allocation has to be addressed. Let $p_i(X, Y, Z)$ be the path loss function at the ith terminal location, for $i = 1, 2, \ldots, a$ where a is the total number of terminals, then

$$p_i(X, Y, Z) = \min_{j=1,\ldots,b} [g_{i,j}(x_j, y_j, z_j)] \qquad (10.17)$$

where $X \equiv (x_1, \ldots, x_b)$; $Y \equiv (y_1, \ldots, y_b)$; $Z \equiv (z_1, \ldots, z_b)$; b is the total number of base-stations; $g_{i,j}(x_j, y_j, z_j)$ is the path loss at the ith terminal location for the base-station located at (x_j, y_j, z_j), computed as Eqn. (10.16).

In such a case, each terminal i is allocated to base-station $\arg\min_j \{g_{i,j}(x_j, y_j, z_j)\}$. A set R_j is hence defined for the set of terminals allocated to the base-station j, where $\cup_{j=1}^{b} R_j = \{1, \ldots, a\}$ and $R_{j_1} \cap R_{j_2} = \emptyset$, $\forall j_1 \neq j_2$.

Constraints. The above formulation is constrained with two basic conditions:

1. The locations of the base-stations are restricted to certain acceptable subsets of the design space; and
2. The maximum power loss at each terminal location over the design space must not exceed a given threshold value.

Objectives. Given a set of terminal layouts, the design objectives of the WLAN can be listed as follows:

1. to minimize the required number of base-stations in order to minimize the cost of the overall system;

2. *Minisum*: to minimize the sum of the path loss predictions over the design space with respect to the base-station location; and
3. *Minimax*: to minimize the maximum of path loss predictions over the coverage space. This function concentrates on the worst case scenario as this may provide the necessary information on maximum path loss in the WLAN.

10.3.2 Multiobjective HGA Approach

In conventional design methodology, it is not only necessary to have a pre-defined number of base-stations, but the conflicting multiobjective functions cannot be solved without the aggregation of the objective functions [197], according to a certain utility function. In many cases, however, the utility function is not well understood prior to the optimization process. Instead of using the conventional heuristic approaches for solving this highly constrained, multiobjective problem, an HGA approach is proposed.

The details of HGA can be referred to in [139]. The main difference lies in the chromosome formulation, as demonstrated in Fig. 10.14 showing the chromosome of base-station location problem.

There are two types of genes, known as control genes and parameter genes. The control genes in the form of bits decide the activation or deactivation of the corresponding base-station. The parameter genes define the x, y, z-coordinates of the base-station locations. For example, in Fig. 10.14, the base-station location (x_1, y_1, z_1), with control gene signified as "0" in the corresponding site, is not being activated. T is the maximum allowable number of base-stations.

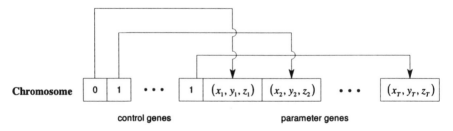

Fig. 10.14. Hierarchical genetic chromosome structure for WLAN

Multiobjective Approach. Based on the constraints and the objectives in the design, it is possible to construct four different objective functions as follows:

10.3 Wireless Local Area Network Design

1. The number of terminals with their path loss higher than the corresponding threshold s_j:

$$f_1 = \sum_{i=1}^{a} q_i \qquad (10.18)$$

where a is the total number of terminals; and

$$q_i = \begin{cases} 1 & \text{if } p_i(X,Y,Z) > s_i \\ 0 & \text{else} \end{cases}$$

and s_i is the specified threshold for the maximum path loss of ith terminal.

2. The number of base-stations required:

$$f_2 = \sum_{i=1}^{T} c_i \qquad (10.19)$$

where c_i is ith bit value in the control genes.

3. The mean of the path loss predictions of the terminals in the design space

$$f_3 = \frac{1}{a} \sum_{i=1}^{a} p_i(X,Y,Z) \qquad (10.20)$$

where $p_i(X,Y,Z)$ is computed as in Eqn. (10.17).

4. The mean of the maximum path loss predictions of the terminals in each set R_j.

$$f_4 = \frac{1}{f_2} \sum_{j=1}^{T} \max_{i \in R_j, c_j=1} p_i(X,Y,Z) \qquad (10.21)$$

All functions are minimized to give a satisfactory design. These four individual objective functions may not be minimized simultaneously in the optimization process. To quantify the available chromosomes, some ranking schemes are required. Consider the following two individual chromosomes I_1 and I_2 with objective values f_1, f_2, f_3, f_4 and f'_1, f'_2, f'_3, f'_4, respectively, I_1 is preferable to I_2 if and only if

Condition I:

$$f_1 < f'_1$$

Condition II:

$$f_1 = f'_1 \text{ and}$$
$$\forall\, i = 2,3,4, \quad f_i \leq f'_i \text{ and}$$
$$\exists\, j = 2,3,4, \quad f_j < f'_j$$

The ranking can thus be made on the fitness assignment of the chromosome and the procedure is described as follows:

1. Sort the population according to the above ranking scheme;
2. Assign the fitnesses of the chromosomes by interpolating the rank from the highest to the worst, using a function:

$$h(I) = h_1 + (h_2 - h_1) \cdot \frac{rank(I) - 1}{N_{pop} - 1} \qquad (10.22)$$

where $rank(I)$ is the rank position of chromosome I in the ordered population, $h(I)$ is the fitness assigned to chromosome I, h_1 and h_2 are the lower and upper limit of fitness respectively, and N_{pop} is the population size;
3. Averaging the fitnesses of the chromosomes in the same rank, so that all of them will be selected at an equal probability.

The advantage of this multiobjective approach is multifold. Firstly, there is no need to determine the penalty factor, which may affect the searching process. Secondly, no combination of objective functions is required. The designer will obtain a Pareto set of the solutions in which any single set of solution can be freely chosen according to the fulfillment of the system requirements. Furthermore, the primary interest, that none of the terminals is higher than the power loss threshold, is also reflected.

Genetic Operations. Since there are two different types of genes, represented in binary and real numbers, the genetic operations should be modified appropriately. For a binary coded control gene, the conventional one-point crossover and bit mutation described in the previous section are used. For a real number coded parameter gene, the specialized genetic operations (random mutations and arithmetic crossovers) stated in [149] are adopted.

10.3.3 Implementation

The Template for the WLAN problem is the same as the conventional GA as shown in Table. 10.17. In order to incorporate the multiobjective ranking scheme, some modifications are undertaken to obtain the parameters and the class, as tabulated in Table 10.3.3.

10.3.4 Results

Without complicating the calculation further, and so obscuring the essence of the HGA design approach, only a 2-D model of a floor plan is proposed for installing the WLAN as depicted in Fig. 10.15. Its dimensions are 75m × 30m. This assumes that there are two different types of wall with $n_1 = 6.0$ and $n_2 = 3.0$, shown as lines of different thickness. There are a total of 200

10.3 Wireless Local Area Network Design

Table 10.17. A template for typical GA application

```
void main()
{
double xRate = 0.01 ;                          // crossover Rate
double mRate = 0.02 ;                          // mutation Rate
int Population_Size = 20 ;                     // Population size
int Sub_Population_Size = 20 ;                 // Sub Population size

Population Pop(Population_Size) ;
Population SubPop(Sub_Population_Size) ;
Genetic Algorithm GAs(Pop,xRate,mRate) ;
GAs.Evaluate(Pop) ;
do
  {
      GAs.Select_Parent(Pop,SubPop) ;          // selection process
      GAs.Recombine(SubPop) ;                  // recombine the 'genes' from sub-population
      GAs.Mutate(SubPop) ;                     // mutate the population statistically
      GAs.Evaluate(SubPop) ;                   // evaluate the fitness value(s) of
                                               // individuals of sub-population

      GAs.Reinsert(Pop,SubPop) ;
      GAs.Generation() ;
  } While(GAs.Terminate = TRUE ) ;
} // End of Main
```

Class   Chromosome
         DEFAULT_SIZE = number of base station
         DEFAULT_CTL = 3   // 3D space parameter control
                                              // for each base station

Class   GeneticAlgorithm
         void GeneticAlgorithm::Evaluate(Population Pop)
         {
           double f1[],f2[],f3[],f4[] ;
           double rank[] ;   // Ranking value of f_1, f_2, f_3, f_4
           Population TempPop(Pop.getSize()) ;
           Calculate_Constrains(f1,Pop) ;   // Eqn. 10.18
           Calculate_Constrains(f2,Pop) ;   // Eqn. 10.19
           Calculate_Constrains(f3,Pop) ;   // Eqn. 10.20
           Calculate_Constrains(f4,Pop) ;   // Eqn. 10.21
           Ranking_Scheme(rank,f1,f2,f3,f4) ;
           // Assign the ranking value according to f_1, f_2, f_3 and f_4
           Update_Fitness(Pop,rank) ;   // Eqn. 10.22
           // using the ranking value to calculate the fitness value of individual in Pop
         }

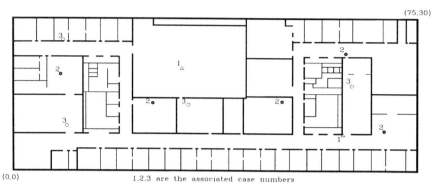

Fig. 10.15. Location of base-stations for 3 different cases

terminals evenly distributed in the area. The location of each terminal is arranged in x-y coordinates of the floor plan which can be computed as

$$\forall\, i = 1, \ldots, 200,$$
$$x_r(i) = 3.75 \cdot ((i-1) \text{ div } 10) + 1.875$$
$$y_r(i) = 3.0 \cdot ((i-1) \bmod 10) + 1.5$$

where $(a \text{ div } b)$ and $(a \bmod b)$ give the quotient and the remainder when a is divided by b, respectively.

Based on the loss model described in Eqn. (10.16), the propagation prediction procedure counts the number of walls which blocked the line of sight between the base-station and the given terminals.

In this arrangement, the chromosome in the form depicted in Fig. 10.14 is applied with the initial arbitrarily chosen maximum number of base-station $T = 8$. The population size of the GA is limited to 100 and 30 offspring are generated in each cycle.

To demonstrate the effectiveness of this design approach, three different power loss requirements are set as tabulated in Table 10.18. This method of HGA approach is required to fulfil all the specifications. After 1000 cycles, the HGA operation is terminated in each case for a direct comparison. The results are shown in Fig. 10.15. The derived base-stations from the HGA are identified as denoted by the case number, and Table 10.19 summarizes the achievable objective values of each case.

From this table, it is clearly demonstrated that the HGA approach is capable of simultaneously identifying the required number of base-stations as well as their corresponding locations. If the power loss threshold requirement

s_i is changed, from 100dB in case 1 to 80dB in case 2, the HGA is capable of dealing with the higher number (five) of base-stations required. This is achieved without further changes to the basic structure of the chromosome.

To further demonstrate the versatility of HGA design, the first 30 terminals which are located on the left hand side ($i = 1,\ldots 30$) are required to reach a higher system requirement as indicated in case 3. As a result, the locations of the base-stations are shifted towards the region in order to meet the design objective as shown in Fig. 10.15.

Table 10.18. Power loss threshold specifications for different cases

Case	Threshold (s_i)
1	100dB
2	80dB
3	70dB ($i \leq 30$) & 100dB ($i > 30$)

Table 10.19. Achievable performance

Case	f_1	f_2	f_3	f_4
1	0	2	69.59dB	91.83dB
2	0	5	61.42dB	74.64dB
3	0	4	64.29dB	79.77dB

An added feature of this design approach is the trade off between cost (number of base-stations) and performance on the basis of minisum and minimax objective functions, f_3 and f_4. It should be noted that this is only possible when the condition of $f_1 = 0$ is reached. A complete range of power losses set against the number of base-stations for each case is identified and these are shown in Fig. 10.16–10.18. This set of results provides a full picture of the WLAN design which offers various alternatives for the final implementation.

314   10. Genetic Algorithms in Communication Systems

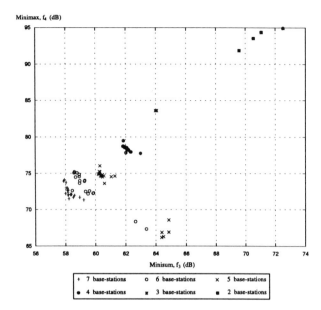

Fig. 10.16. Final population for case 1

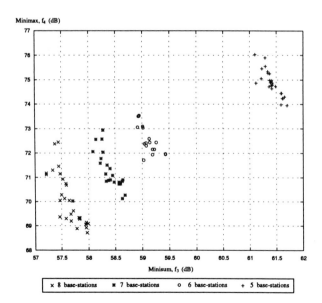

Fig. 10.17. Final population for case 2

10.3 Wireless Local Area Network Design 315

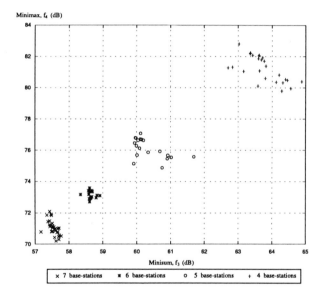

Fig. 10.18. Final population for case 3

Appendix A

Least Mean Square Time Delay Estimation (LMSTDE)

The time delay is estimated by a Finite Impulse Response (FIR) model. The estimation error $e(k)$ is then equal to

$$\begin{aligned} e(k) &= y(k) - AX(k) \\ &= y(k) - \sum_{i=-L}^{L} a_i x(k-i) \end{aligned} \quad (A.1)$$

where $A = \begin{bmatrix} a_{-L} & a_{-L+1} & \cdots & a_L \end{bmatrix}$ is the $(2L+1)$ filter parameter vector; $X(k) = \begin{bmatrix} x(k+L) & x(k+L-1) & \cdots & x(k-L) \end{bmatrix}^T$ is the input vector; and $y(k)$ is the delayed signal.

The filter weights a_i are updated by minimizing the mean square error (MSE) as below:

$$A(k+1) = A(k) + 2\mu_w e(k) X^T(k) \quad (A.2)$$

where μ_w is the gain constant that regulates the speed and stability of adaptation.

Appendix B

Constrained LMS Algorithm

The constrained LMS algorithm is formulated as follows:

$$\begin{aligned}\hat{D}(k+1) &= \hat{D}(k) - \mu_n \frac{\partial e^2(k)}{\partial \hat{D}} \\ &= \hat{D}(k) - \mu_n e(k) \sum_{n=-L}^{L} x(k-n) f(n - \hat{D}(k))\end{aligned} \quad (\text{B.1})$$

where

$$e(k) = y(k) - \sum_{n=-L}^{L} sinc(n - \hat{D}(k)) x(k-n)$$

$$f(v) = \frac{\cos(\pi v) - sinc(\pi v)}{v}$$

and μ_n is a convergence factor controlling the stability.

The initial value of $\hat{D}(0)$ must be within the range of $D \pm 1$ so as to retain a unimodal error surface of $e(k)$.

Appendix C

Linear Distributed Random Generator

According to the random mutation expressed in Eqn. 2.6, a Gaussian distributed random number is added on the genes for mutation. Such a random process is not easy to generate in hardware. However, the design of a pseudo random number generator is simple but possesses a uniform distribution that is not suitable for this application. Therefore, a new method to generate the approximated Gaussian distributed random numbers has been proposed by simply manipulating the pseudo random number formulation.

Considering that the output of the random function Ψ in Eqn. 2.6 is formulated as

$$\Psi = \begin{cases} b & \text{if } a > b \\ 0 & \text{else} \end{cases} \quad (C.1)$$

where a, b in this case, are the outputs of two independent pseudo random generators, with the output of each pseudo random generator being set to $(\mu - 3\sigma, \mu + 3\sigma)$. The distribution of the pseudo random number generated is indicated in Fig. C.1(b).

In this way, an approximated Gaussian distribution was obtained as shown in Fig. C.1(c). This distribution was found to be realistic and very similar to that obtained from a true Gaussian noise generator as shown in Fig. C.1(a).

322   Appendix C

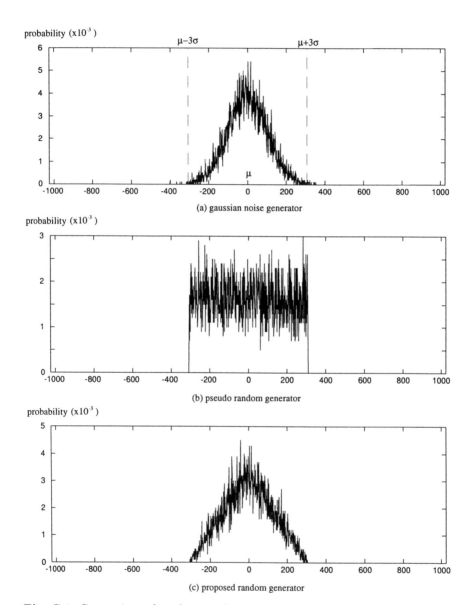

Fig. C.1. Comparison of random number generators

Appendix D

Multiplication Algorithm

A high-speed VLSI multiplication algorithm using redundant binary representation was implemented, and signed digit number representation [7] was adopted. This representation was a fixed radix 2 and a digit set $\{\bar{1}, 0, 1\}$ where $\bar{1}$ denotes -1. An n-digits redundant binary integer $Y = [y_{n-1} \cdots y_0]_{SD2} (y_i \in \{\bar{1}, 0, 1\})$ has the value $\sum_{i=0}^{n-1} y_i \times 2^i$.

The multiplier algorithm based on the redundant binary representation [213] is formed by a binary tree of redundant binary adders. Multicand and multiplier are converted into equivalent redundant binary integers and then, an amount of n n-digit partial products represented in redundant binary representation is generated. The computations can be performed in a constant time independent of n.

The partial products were added up in pairs by means of a binary tree of redundant binary adders and the product represented in the redundant binary representation was obtained. The addition of two numbers in the redundant binary number system can be carried out in a constant time independent of the word length of operands. The constant time addition of two redundant binary numbers can be realized by the Carry-Propagation-Free Addition (CPFA). The CPFA is performed in two steps:

1. it is used to determine the intermediate carry $c_i (\in \{\bar{1}, 0, 1\})$ and the intermediate sum digit $s_i (\in \{\bar{1}, 0, 1\})$ at each position, which satisfies the equation $x_i + y_i = 2c_i + s_i$, where x_i and y_i are the augend and addend digits. There are six types of combinations of the two digits in addition as tabulated in Table D.1; and
2. the sum digit $z_i (\in \{\bar{1}, 0, 1\})$ at each position is obtained by adding the intermediate sum digit s_i and the intermediate carry c_{i-1} from the next-lower-order position, without generating a carry at any position in the second step. As a result, the additions are performed in a time proportional to $log_2 n$.

Table D.1. Computation rules For CPFA

Type	Combination $\{x_i, y_i\}$	next-lower-order-position $\{x_{i-1}, y_{i-1}\}$	carry c_i	sum s_i
1	$\{1, 1\}$	—	1	0
2	$\{1, 0\}$	Both are nonnegative	1	$\bar{1}$
		Otherwise	0	1
3	$\{1, \bar{1}\}$	—	0	0
4	$\{0, 0\}$	—	0	0
5	$\{0, \bar{1}\}$	Both are nonnegative	0	$\bar{1}$
		Otherwise	$\bar{1}$	1
6	$\{\bar{1}, 1\}$	—	$\bar{1}$	0

Finally, the product must be converted into binary representation. As an n-digit redundant binary number

$$A \left(= \sum_{i=0}^{n-1} a_i \times 2^i, a_i \in \{\bar{1}, 0, 1\} \right)$$

is equal to

$$A^+ \left(= \sum_{a_i = 1} a_i \times 2^i \right) - A^- \left(= \sum_{a_i = \bar{1}} (-a_i) \times 2^i \right)$$

Therefore, a conversion of an n-digit redundant binary integer into the equivalent $(n+1)$-bit 2's complement binary integer is performed by subtracting A^- from A^+, where A^- and A^+ are n bit unsigned binary integers from the positive digits and the negative digits in A, respectively. This conversion can be performed in a time proportional to $log_2 n$ by means of an ordinary carry-look-ahead adder. In addition, the extended Booth's algorithm can be applied to further reduce the computation time and the amount of hardware required.

Appendix E

Digital IIR Filter Designs

Based on the HGA formulation as indicated in Chap. 5.1, the design of digital filters in the form of LP, HP, BP and BS are thus possible. The genetic operational parameters are shown in Tables E.1 and E.2

Table E.1. Parameters for genetic operations

Population Size	100
Generation Gap	0.2
Selection	Multiobjective Ranking
Reinsertion	Replace the lowest Rank
sflag = 1	$f_1 = 0 \wedge f_2 = 0$
N_{max}	20000

Table E.2. Parameters for chromosome operations

	Control Genes	Coefficient Genes
Representation	Bit Representation (1 bit)	Real Number Representation
Crossover	Normal Crossover	Normal Crossover
Crossover Rate	0.85	0.8
Mutation	Bit Mutation	Random Mutation
Mutation Rate	0.15	0.1

The fundamental structure of $H(z)$ which applies to all four filters is given as:

$$H(z) = K \prod_{i=1}^{3} \frac{(z + b_i)}{(z + a_i)} \prod_{j=1}^{4} \frac{(z^2 + b_{j1}z + b_{j2})}{(z^2 + a_{j1}z + a_{j2})} \quad (E.1)$$

The control genes (g_c) and coefficient genes (g_r) in this case are thus

$$g_c \in B^{14} \quad (E.2)$$

$$g_r = \left\{ \begin{array}{l} a_1, a_2, a_3, a_{11}, a_{12}, a_{21}, a_{22}, a_{31}, a_{32}, a_{41}, a_{42}, \\ b_1, b_2, b_3, b_{11}, b_{12}, b_{21}, b_{22}, b_{31}, b_{32}, b_{41}, b_{42} \end{array} \right\} \quad \text{(E.3)}$$

where $B = [0, 1]$ and the ranges of $a_i, b_i, a_{i1}, a_{i2}, b_{i1}, b_{i2}$ are defined as in Table 5.2.

The design criteria for the filters is tabulated in Table E.3.

Table E.3. Summary of filter performances

Filter Type	Design Criteria		K
LP	$0.89125 \leq \left|H(e^{j\omega})\right| \leq 1,$ $\left|H(e^{j\omega})\right| \leq 0.17783,$	$0 \leq |\omega| \leq 0.2\pi$ $0.3\pi \leq |\omega| \leq \pi$	$|H(1)| = 1$
HP	$\left|H(e^{j\omega})\right| \leq 0.17783,$ $0.89125 \leq \left|H(e^{j\omega})\right| \leq 1,$	$0 \leq |\omega| \leq 0.7\pi$ $0.8\pi \leq |\omega| \leq \pi$	$\left|H(e^{j\pi})\right| = 1$
BP	$\left|H(e^{j\omega})\right| \leq 0.17783,$ $0.89125 \leq \left|H(e^{j\omega})\right| \leq 1,$	$0 \leq |\omega| \leq 0.25\pi$ $0.75\pi \leq |\omega| \leq \pi$ $0.4\pi \leq |\omega| \leq 0.6\pi$	$\left|H(e^{0.5\pi j})\right| = 1$
BS	$\left|H(e^{j\omega})\right| \leq 0.17783,$ $0.89125 \leq \left|H(e^{j\omega})\right| \leq 1,$	$0.4\pi \leq |\omega| \leq 0.6\pi$ $0 \leq |\omega| \leq 0.25\pi$ $0.75\pi \leq |\omega| \leq \pi$	$|H(1)| = 1$

Appendix F

Development Tools

The study and evaluation of GA, are essentially non-analytic, largely depending on simulation. While they are strongly application independent, GA software has potentially a very broad domain of application. Part of the common software package is briefly introduced and more information can be found in [98].

Genetic Algorithm Toolbox in MATLAB

A GA Toolbox is developed [33] for MATLAB [146]. Given the versatility of MATLAB's high-level language, problems can be coded in m-files easily. Coupling this with MATLAB's advanced data analysis, visual tools and special purpose application domain toolboxes, the user is presented with a uniform environment with which to explore the potential of GA.

GENESIS

GENEtic Search Implementation System (GENESIS) was developed by John Grefenstette [90]. It is a function optimization system based on genetic search techniques. As the first widely available GA programme, GENESIS has been very influential in stimulating the use of GA, and several other GA packages have been generated because of its capability.

A real number representation and binary representation are allowable. A number of new options have been added, including: a display mode that includes an interactive user interface, the option to maximize or minimize the objective function, the choice of rank-based or proportional selection algorithm, and an option to use a Gray code as a transparent lower level representation.

GENOCOP

GEnetic Algorithm for Numerical Optimization for COnstrained Problems (GENOCOP) was developed by Zbigniew Michalewicz and details can be obtained in [149]. The GENOCOP system has been designed to find a global optimum (minimum or maximum) of a function with additional linear equalities and inequalities constraints. It runs on any UNIX and DOS system.

GENEsYs

GENEsYs [224] is a GENESIS-based GA implementation which includes extensions and new features for experimental purposes. Different Selection schemes like linear ranking, Boltzmann, (μ, λ)-selection, and general extinctive selection variants are included. Crossover operators and self-adaptation of mutation rates are also possible. There are additional data-monitoring facilities such as recording average, variance and skew of object variables and mutation rates, and creating bitmap-dumps of the population.

TOLKIEN

TOLKIEN (TOoLKIt for gENetics-based applications) ver 1.1 [222] is a C++ class library named in memory of J.R.R. Tolkien. A collection of reusable objects have been developed for genetics-based applications. For portability, no compiler specific or class library specific features are used. The current version has been compiled successfully using Borland C++ Ver. 3.1 and GNU C++. TOLKIEN contains a number of useful extensions to the generic GA. For example:

– chromosomes of user-definable types; binary, character, integer and floating point chromosomes are provided;
– gray code encoding and decoding;
– multi-point and uniform crossover;
– diploidy;
– various selection schemes such as tournament selection and linear ranking
– linear fitness scaling and sigma truncation.

Distributed GENESIS 1.0

Distributed GENESIS 1.0 (DGENESIS) was developed by Erick Cantú-Paz. It is an implementation of migration genetic algorithms, described in Sect. 3.1.2. Its code is based on GENESIS 5.0. Each subpopulation is handled by a UNIX process and communication them is handled with Berkeley sockets.

The user can set the migration rate, migration interval and the topology of the communication between subpopulations in order to realize migration GA.

This version of DGENESIS requires the socket interface provided with 4.2BSD UNIX. It has run successfully on DECStations (running Ultrix 4.2), Sun workstations (with SunOS), microVAXes (running Ultrix 4.1) and PCs (with 386BSD 0.1).

In any network, there are fast and also slow machines. To make the most of available resources, the work load in the participating systems can be balanced by assign to each machine a different number of processes according to their capabilities.

Generic Evolutionary Toolbox

Generic Evolutionary Toolbox (GenET) is a generic evolutionary algorithm, a toolbox for fast development of GA applications and for research in evaluating different evolutionary models, operators, etc.

The package, in addition to allowing for fast implementation of applications and being a natural tool for comparing different models and strategies, is intended to become a depository of representations and operators. Currently, only floating point representation is implemented in the library with few operators.

The algorithm provides a wide selection of models and choices. For example, POPULATION models range from generational GA, through steady-state, to (n,m)-EP and (n,n+m)-EP models (for arbitrary problems, not just parameter OPTIMIZATION). (Some are not finished at the moment). Choices include automatic adaptation of operator probabilities and a dynamic ranking mechanism, etc.

References

1. Actel Corporation (1994): FPGA Data Book and Design Guide.
2. Alidaee B. (1994): Minimizing absolute and squared deviation of completion times from due dates. Production and Operations Management, **2(2)**, 133–147.
3. Anderson, E.J. and M.C. Ferris (1990): A genetic Algorithm for the assembly line balancing problem. Technical Report TR 926, Computer Sciences Department, University of Wisconsin-Madison.
4. Angeline, P.J., G.M. Saunders, and J.B. Pollack (1994): An evolutionary algorithm that constructs recurrent neural networks. IEEE Trans. Neural Networks, **5(1)**, 54–65.
5. Asakawa, K and Hideyuki Takagi (1994): Neural networks in Japan. Communication of the ACM, **37(3)**, 106–112.
6. Atal, B.S. (1974): Effectiveness of linear prediction characteristics of speech wave for automatic speaker identification and Verification. Journal of the Acoustic Society of Amer., **55**, 1304–1312.
7. A. Avizienis (1961): Signed-digit number representations for fast parallel arithmetic., IEEE Trans. Electron. Comput. **EC-10**, 389–400.
8. Bahl, L.R., J.K. Baker, P.S. Cohen, A.G. Cole, F. Jelinek, B.L. Lewis, and R.L. Mercer (1978): Automatic recognition of continuously spoken sentences from a finite state grammar. Proc. ICASSP, 418–421.
9. Baker, J.E. (1985): Adaptive selection methods for genetic algorithms. Proc. 1st Int Conf on Genetic Algorithms, 101–111.
10. Baker J.E. (1987): Reducing bias and inefficiency in the selection algorithms. Proc. 2nd Int. Conf. Genetic Algorithms. Lawrence Erlbaum Associates, Hillsdale, NJ, 14–21.
11. Baker, J.K. (1975): The DRAGON system - An overview. IEEE Trans. Acoustics, Speech and Signal Processing, **ASSP-23**, Feb, 24–29.
12. Baker K.R. and G. D. Scudder (1990): Sequencing with earliness and tardiness penalties: A review. Operation Research, **38**, 22–36.
13. Bakis, R. (1976): Continuous speech word recognition via centisecond acoustic states. Proc. of ASA Meeting (Washington DC).
14. Baluja, S. (1993): Structure and performance of fine-grain parallelism in genetic search. Proc. 5th Int. Conf. Genetic Algorithm.
15. Baum, L.E. and J.A. Egon (1967): An inequality with applications to statistical estimation for probabilistic functions of a markov process and to a model for ecology. Bull. Amer. Meteorol. Soc. **(73)**, 360-363.
16. Baum, L.E., T. Petrie, G. Soules, and N. Weiss (1970): A maximization technique occurring in the statistical analysis of probabilistic functions of markov chains. Ann. Math. Stat. **41(1)**, 164–171.

17. Baum, L.E. (1972): An inequality and associated maximization technique in statistical estimation for probabilistic functions of markov process. Inequalities III, 1–8.
18. Beasley, D., D.R. Bull, R.R. Martin (1993): An overview of genetic algorithms: Part 1, fundamentals. University Computing, **15(2)**, 58–69.
19. Beasley, D., D.R. Bull, R.R. Martin (1993): An overview of genetic algorithms: Part 2, research topics. University Computing, **15(4)**, 170–181.
20. Berge, O., K.O. Petterson and S. Sorzdal (1988): Active cancellation of transformer noise: Field measurements. Applied Acoustics, **23**, 309–320.
21. D. Bertsekas and R. Gallager (1992): Data networks. 2nd ed., Prentice-Hall, 1992.
22. Booker, L. (1987): Improving search in genetic algorithms. Genetic Algorithms and Stimulated Annealing, L. Davis (Eds), 61–73.
23. Braun, H. (1990): On solving travelling salesman problems by genetic algorithms. Proc. First Workshop Parallel Problem Solving from Nature, Springer Verlag, Berlin, 129–133.
24. Brown, K. and et al. (1984): The DTWP: An LPC based dynamic time warping processor for isolated word recognition. Bell System Tech. Journal, vol. **63(3)**.
25. Cantú-Paz, E. (1995): A summary of research on parallel genetic algorithms. IlliGAL Report No. 95007, Illinois Genetic Algorithms Laboratory, University of Illinois at Urbana-Champaign.
26. Chan, Y.T., J.M. Riley, and J.B. Plant (1981): Modeling of time delay and its application to estimation of nonstationary delays. IEEE Trans. Acoust., Speech, Signal Processing, **ASSP-29**, 577–581.
27. Chang, C. and B. H. Juang (1992): Discriminative template training of dynamic programming based speech recognizers. Proc ICASSP, ICASSP-92, San Francisco, March.
28. Chau, C.W., S. Kwong, C.K. Diu, and W.R. Fahrner (1997): Optimisation of HMM by a genetic algorithm. Proc. ICASSP, 1727–1730.
29. Chau, C.W. (1997): A multiprocessor system for speech recognition. Master of Philosophy, City University of Hong Kong, October.
30. Chen, D., C. Giles, G. Sun, H. Chen, Y. Less, and M. Goudreau (1993): Constructive learning of recurrent neural network. Proc. IEEE Int. Conf. Neural Network **3**, 1196–1201.
31. Cheng T.C.E. and S. Podolsky (1993): Just-in-Time Manufacturing-An Introduction. Chapman & Hall.
32. Cheuk, K.P., K.F. Man, Y.C. Ho and K.S. Tang (1994): Active noise control for power transformer. Proc. Inter-Noise 94, 1365–1368.
33. Chipperfield, A.J., P.J. Fleming and H. Pohlheim (1994): A genetic algorithm toolbox for MATLAB. Proc. Int. Conf. on Systems Engineering, Coventry, UK, 6–8.
34. Chipperfield, A.J. and P.J. Fleming (1994): Parallel genetic algorithms: A survey. ACSE Research Report No. 518, University of Sheffield.
35. Cobb, H.G. (1990): An investigation into the use of hypermutation as an adaptive operator in genetic algorithms having continuous, time-dependent nonstationary environments. NRL Memorandum Report 6760.
36. Cobb, H.G. and J.J. Grefenstette (1993): Genetic algorithms for tracking changing environments. Proc. 5th Int. Conf. Genetic Algorithms, 523–530.
37. Cohoon, J.P., W.N. Martin and D.S. Richards (1991): A multi-population genetic algorithm for solving the k-partition problem on hyper-cubes. Proc. 4th Int. Conf. Genetic Algorithms, 244–248.

38. Cooley, J.W. and J. W. Tukey (1965) An algorithm for the machine computation of complex Fourier series. Math Computation, **19**, 297–381.
39. Daniels R.W. (1974): Approximation methods for electronic filter design. McGraw-Hill Book Company, NY.
40. Dautrich, B.A., L.R. Rabiner and T.B. Martin (1983): The effect of selected signal processing techniques on the performance of a filter based isolated word recognizer. Bell System Tech. J., **62(5)**, 1311–1336, May-June.
41. B. A. Dautrich, L.R. Rabiner and T.B. Martin, (1983): On the effects of varying filter bank parameters on isolated word recognition. IEEE Trans. Acoustic, Speech and Signal Processing, **ASSP-31(4)**, 793–807.
42. Davidor, Y. (1991): A genetic algorithm applied to robot trajectory generation. Handbook of Genetic Algorithms, L. Davis (Eds), 144–165.
43. Davies, R. and T. Clarke (1995): Parallel implementation of a genetic algorithm. Control Eng. Practice, **3(1)**, 11–19.
44. Davis J.S. and J. Kanet (1993): Single-machine scheduling with early and tardy completion costs. Naval Research Logistics, **40**, 85–101.
45. Davis, L. (1985): Job shop scheduling with genetic algorithms. Proc. 1st Int. Conf. Genetic Algorithms, J.J. Grefenstette (Eds), 136–140.
46. Davis, L. (1989): Adapting operator probabilities in genetic algorithms. Proc. 3rd Int. Conf. Genetic Algorithms, 61–69.
47. Davis, L. (1991): Handbook of genetic algorithms. Van Nostrand Reinhold.
48. Deb K. and D.E. Goldberg (1991): Analyzing deception in trap functions. Technical Report IlliGAL 91009, Department of Computer Science, University of Illinois at Urbana-Champaign, Urbana.
49. DeJong, K. (1975): The analysis and behaviour of a class of genetic adaptive systems. PhD thesis, University of Michigan.
50. DeJong, K.A. and W.M. Spears (1990) : An analysis of the interacting roles of population size and crossover in genetic algorithms. Proc. First Workshop Parallel Problem Solving from Nature, Springer Verlag, Berlin, 38–47.
51. De P., J.B. Jhosh and C.E. Wells (1991), Scheduling to minimize weighted earliness and tardiness about a common due-date. Computer Operation Research, **18(5)**, 465–475.
52. De P., J.B. Jhosh and C.E. Wells (1993), On general solution for a class of early/tardy problems. Computer Operation Research, **20**, 141–149.
53. Dodd, N., D. Macfarlane and C. Marland (1991): Optimization of artificial neural network structure using genetic techniques implemented on multiple transputers. Transputing '91, **2** 687–700.
54. Dynan, W.S. and R. Tjian (1985): Control of eukaryotic messenger RNA synthesis by sequence-specific DNA-binding proteins. Nature **316**, 774–778.
55. R. Elbaum and M. Sidi (1995): Topological design of local area networks using genetic algorithms. *IEEE Infocom95*, 1c.1.1–1c.1.8.
56. Elliott, S.J., P.A. Nelson, I.M. Stothers and C.C. Boucher (1990): In-flight experiments on the active control of propeller-induced cabin noise. J. Sound and Vibration, **140**, 219–238.
57. Elliott, S.J. and P.A. Nelson, (1993): Active noise control. IEEE Signal Processing Magazine, Oct, 12–35.
58. Eriksson, L.J. (1991): Development of the filtered-U algorithm for active noise control. J. Acoustic Soc. Am **89**, 257–265.
59. Eshelman, L.J., R. Caruna, and J.D. Schaffer (1989): Biases in the crossover landscape. Proc. 3rd Int. Conf. Genetic Algorithms, 10–19.
60. Fitzpatrick, J.M. and J.J. Grefenstette (1988): Genetic algorithms in noisy environments. Machine Learning, **3(2/3)**, 101–120.

61. Flapper S.D.P., G.J. Miltenburg and J. Wijngaard (1991): Embedding JIT into MRP. International Journal of Production Research, **29(2)**, 329–341.
62. Fonseca, C.M., E.M. Mendes, P.J. Fleming and S.A. Billings (1993): Non-linear model term selection with genetic algorithms. Proc. Workshop on Natural Algorithms in Signal Processing, 27/1–27/8.
63. Fonseca, C.M. and P.J. Fleming (1993): Genetic algorithms for multiobjecitve optimization: formulation, discussion and generalization. Proc. 5th Int. Conf. Genetic Algorithms, (S. Forrest, ed.), 416–423.
64. Fonseca, C.M. and P.J. Fleming (1994): An overview of evolutionary algorithms in multiobjective optimization. Research Report No. 527, Dept. of Automatic Control and Systems Eng., University of Sheffield, UK.
65. Fonseca, C.M. and P.J. Fleming (1995): Multiobjecitve genetic algorithms made easy: selection, sharing and mating restriction. Proc. 1st IEE/IEEE Int. Conf. on GAs in Engineering Systems: Innovations and Applications, 45–52.
66. Fourman, M.P. (1985): Compaction of symbolic layout using genetic algorithm. Proc. 1nd Int. Conf. Genetic Algorithms, 141–153.
67. Fu, L.M. (1994): Neural networks in computer intelligence. McGraw-Hill.
68. Furui, S. (1986): Speaker-independent isolated word recognition using Dynamic Features of Speech Spectrum. IEEE Trans. acoustic, speech and signal processing, **ASSP-34(1)**.
69. Furui, S. (1988): A VQ-based preprocessor using cepstral dynamic features for speaker-independent large vocabulary word recognition. IEEE Trans. acoustic, speech and signal processing, **ASSP-36(7)**.
70. Garey, M.R. and D.S. Johnson (1979): Computers and intractability: a guide to the theory of NP-completeness. Freeman, San Francisco.
71. Gerla, M. and L. Kleinrock (1977): On the topological design of distributed computer networks. IEEE Trans on Commun., **COM-25(1)**, 48–60.
72. Gerla, M., J.A.S. Monteiro and R. Pazos (1989): Topology design and bandwidth allocation in ATM nets. IEEE J. Selected Areas in Communications, **7(8)**, 1253–1262
73. Gessener R.A. (1986): Master production schedule planning. John Wiley & Sons, Inc., New York.
74. Gill, P.E., W. Murray and M.H. Wright (1981): Practical optimization. Academic Press.
75. Gillies, A.M. (1985): Machine learning procedures for generating image domain feature detectors. Doctoral Dissertation, University of Michigan.
76. Glover, K. and D. McFarlane (1989): Robust stabilization of normalized coprime factor plant descriptions with -bounded Uncertainty. IEEE Trans. Automat. Contr., **AC-34(8)**, 821–830.
77. Goldberg, D.E. (1987): Simple genetic algorithms and the minimal deceptive problem. Genetic Algorithms and Stimulated Annealing, L. Davis (Ed.) 74–88.
78. Goldberg, D.E. (1989): Genetic algorithms in search, optimization and machine learning. Addison-Wesley.
79. Goldberg, D.E. (1990): Real-coded genetic algorithms, virtual alphabets, and block. Technical Report No. 90001, University of Illinois.
80. Goldberg, D.E. and R. Lingle (1985): Alleles, locis, and the traveling salesman problem. Proc. Int. Conf. Genetic Algorithms and Their Applications, 154–159.
81. Goldberg, D.E. and J.J. Richardson (1987): Genetic algorithms with sharing for multimodal function optimization. Proc. 2nd Int. Conf. Genetic Algorithms, 41–47.

82. Goldberg, D.E. and R.E. Smith (1987): Nonstationary function optimization using genetic dominance and diploidy. Proc. 2nd Int. Conf. Genetic Algorithms, 59–68.
83. Gordon, V. and D. Whitley (1993): Serial and parallel genetic algorithms as function optimizer. Proc. 5th Int Conf. Genetic Algorithms, 177–183.
84. Gorges-Schleuter, M. (1989): ASPARAGOS An asynchronous parallel genetic optimization strategy. Proc. 3rd Int. Conf. Genetic Algorithms, 422–427.
85. Graebe, S.F. (1994): Robust and Adaptive Control of an Unknown Plant: A Benchmark of New Format. Automatica, **30(4)**, 567–575.
86. Grefenstette, J.J. (1986): Optimization of control parameters for genetic algorithms. IEEE Trans Systems, Man, and Cybernetics, **SMC-16(1)**, 122–128.
87. Grefenstette, J.J. (1992): Genetic algorithms for changing environments. Parallel Problem Solving from Nature, 2, 137–144.
88. Grefenstette, J.J. (1993): Deception considered harmful. Foundations of Algorithms, 2, L. Darrell Whitley (Ed.) 75–91.
89. Grefenstette J.J. and J. Baker (1989): How genetic algorithms work: A critical look at implicit parallelism. Proc 3rd Int. Conf. Genetic Algorithm.
90. Grefenstette J.J. (1990): A user's guide to GENESIS v5.0. Naval Research Laboratory, Washington, D.C.
91. Guillemin, E.A. (1956): Synthesis of passive networks. John Wiley and Sons, NY.
92. Hajela, P. and Lin, C.-Y. (1992): Genetic search strategies in multicriterion optimal design. Structural Optimization, 4, 99–107.
93. Hall, H.R., W.B. Ferren and R.J. Bernhard (1992): Active control of radiated Sound from ducts. Trans. of the ASME, 114, 338–346.
94. Hall N.G. and M.E. Posner (1991): Earliness-tardiness scheduling problem I: Weighted deviation of completion times about a common due date. Operation Research, **39(5)**, 836–846.
95. Hall N.G., W. Kubiak, and S.P. Sethi (1991): Earliness-tardiness scheduling problem II: Deviation of completion times about a restrictive common due date. Operation Research, **39(5)**, 847–856.
96. Hax A.C, and D. Candea (1983), Production and inventory management. Prentice-Hall, New Jerar.
97. Heady R.B. and Z. Zhu (1998): Minimizing the sum of job earliness and tardiness in a multimachine system. International Journal of Production Research, **36(6)**, 1619–1632.
98. Heitkoetter, J. and D. Beasley (Eds) (1994): The Hitch-Hiker's guide to evolutionary computation: A list of frequently asked questions (FAQ). USENET:comp.ai.genetic., 1994.
99. Helms, H.D. (1967): Fast Fourier transform method of computing difference equations and simulating filters. IEEE transactions on Audio and Electroacoustics, **15(2)**, 85–90.
100. Ho C.H. (1989): Evaluation the impact operating environment on MRP system nervousness. International Journal of Production Research, 26, 1–18.
101. Ho, K.C., Y.T. Chan and P.C. Ching (1993): Adaptive time-delay estimation in nonstationary signal and/or noise power environments. IEEE Trans. Signal Processing, **41(7)**, 2289–2299.
102. Ho, Y.C., K.F. Man, K.P. Cheuk and K.T. Ng (1994): A fully automated water supply system for high rise building. Proc. 1st Asian Control Conference, 1–4.
103. Ho, Y.C., K.F. Man, K.S. Tang and C.Y. Chan (1996): A dependable parallel architecture for active noise control. IFAC World Congress 96, 399–404.

104. Hodgson T.J. and D.W. Wang (1991): Optimal push/pull control strategies for a parallel multistage system: part 1. International Journal of Production Research, **29(6)**, 1279–1287.
105. Hodgson T.J. and D.W. Wang (1991): Optimal push/pull control strategies for a parallel multistage system: part 2. International Journal of Production Research, **29(6)**, 1453–1460.
106. Holland, J.H. (1975): Adaption in natural and artificial systems. MIT Press.
107. Hollstien, R.B. (1971): Artificial genetic adaptation in computer control systems. PhD thesis, University of Michigan.
108. Homaifar, A. and Ed McCormick (1995): Simultaneous design of membership functions and rule sets for fuzzy controllers using genetic algorithms. IEEE Trans Fuzzy Systems, **3(2)**, 129–139.
109. Horn, J. and N. Nafpliotis (1993): Multiobjective optimization using the niched pareto genetic algorithm. IlliGAL Report 93005, University of Illinois at Urbana-Champaign, Urbana, Illinois, USA.
110. Hoyle, D.J., R.A. Hyde and D.J.N. Limebeer (1991): An approach to two degree of freedom design. Proc. 30th IEEE Conf. Dec. Contr., 1581–1585.
111. Huson M. and D. Nanda (1995): The impact of just-In-time on firm performance in the US. Journal of Operations Management, **12(3 & 4)**, 297–310.
112. Itakura, F.I. (1975) Minimum prediction residual principle applied to speech recognition. IEEE Trans. Acoustic, Speech and Signal Processing, **ASSP-23**, Feb, 67-72.
113. Jakob, W., M. Gorges-Schleuter and C. Blume (1992): Application of genetic algorithms to task planning and learning. Parallel Problem Solving from Nature, 2, 291–300.
114. Janikow, C.Z. and Z. Michalewicz (1991): An experimental comparison of binary and floating point representations in genetic algorithms. Proc. 4th Int. Conf. Genetic Algorithms, 31–36.
115. Jang, J.-S.R. and C.-T. Sun (1995): Neuro-fuzzy modeling and control. Proc. IEEE, **83(3)**, 378–406.
116. Jelinek, F. (1973): The development of an experimental Discrete Dictation Reconigzer. Proc. IEEE, **73**, 1616–1624.
117. F. Jelinek (1976): Continuous speech recognition by statistical methods, Proc. IEEE, **64**, April, 250–256.
118. Jones, K.A., J.T. Kadonga, D.J. Rosenfeld, T.J. Kelly and R. Tjian (1987): A cellular DNA binding protein that activates eukaryotic transcription and DNA replication. Cell **48**, 79–84.
119. Karr, C.L. (1991): Genetic algorithms for fuzzy controllers. AI Expert, **6(2)**, 26–33.
120. Karr, C.L. and E.J. Gentry (1993): Fuzzy control of pH using genetic algorithms. IEEE Trans Fuzzy Systems **1(1)**, 46–53.
121. Kennedy, S.A. (1991): Five ways to a smarter genetic algorithm. AI Expert, Dec, 35–38.
122. A. Kershenbaum (1993): Telecommunications network design algorithms. McGraw-Hill.
123. Kido, K., M. Abe and H. Kanai (1989): A new arrangement of additional sound source in an active noise control system. Proc. Inter-Noise 89, 483–488.
124. Kim Y.D. and C. A. Yano (1994): Minimizing mean tardiness and earliness in single-machine scheduling problem with unequal due dates. Naval Research Logistics, **41**, 913–933.
125. Kornberg, A. (1980): DNA replication. Freeman, San Francisco.

126. Kröger, B., P. Schwenderling and O. Vornberger (1993): Parallel genetic packing on transputers. Parallel Genetic Algorithms: Theory and Applications, Amsterdam: IOS Press, 151–185.
127. Kwong, S., C.W. Chau and W. A. Halang (1996): Genetic algorithm for optimizing the nonlinear time alignment of speech recognition system. IEEE Trans. Industrial Electronics, vol. **43(5)**, 559–566
128. Kwong, S., Q. He and K.F. Man: Genetic time warping for isolated word recognition. International Journal of Pattern Recognition and Artificial Intelligence.
129. Lam, H.Y.-F. (1979): Analog and digital filters: design and realization. Prentice-Hall, Englewood Cliffs, NJ.
130. S.E. Levinson, L.R. Rabiner, and M.M. Sondhi (1983): An introduction to the application of the theory of probabilistic functions of a markov process to automatic speech recognition. The Bell Sys. Tech. J., April, 1035–1074.
131. Leug, P. (1936): Process of silencing sound oscillations. U.S. Patent No. 2,043,416.
132. Li Y. D.W. Wang and W.H. Ip (1998): Earliness/tardiness production scheduling and planning and solutions, International Journal of Production Planning & Control, **9(3)**, 275–285.
133. Li Y., W.H. Ip and D.W. Wang (1998): Genetic algorithm approach to scheduling and planning problem. International Journal of Production Economics, **54(1)**, 64–74.
134. Limebeer, D.J.N. (1991): The Specification and purpose of a controller design case study. Proc. 30th IEEE Conf. Dec. Contr., Brighton, England, 1579–1580.
135. Louis, S.J. and Rawlins, G.J.E. (1993): Pareto optimality, GA-easiness and deception. Proc. 5th Int. Conf. Genetic Algorithms, 118–223.
136. Mangasarian, O.L., and W.H. Wolberg (1990): Cancer diagnosis via linear programming. SIAM News, **23(5)**, 1 & 18.
137. Mahfoud, S.W. (1992): Crowding and preselection revisited. IlliGAL Report No. 92004, Department of Computer Science, Univeristy of Illinois at Urbana-Champaign.
138. Mahfoud, S.W. (1994): Population sizing for sharing methods. IlliGAL Report No. 94005, Department of Computer Science, University of Illinois at Urbana-Champaign, Urbana.
139. K.F. Man, K.S. Tang, S. Kwong and W.A. Halang (1997): Genetic algorithms for control and signal processing. Springer-Verlag, ISBN 3-540-76101-2.
140. Man, K.F., K.S. Tang and S. Kwong (1996): Genetic algorithms: concept and applications. IEEE Trans. Industrial Electronics **43(5)**, 519–534.
141. Manderick, B. and P. Spiessens (1989): Fine-grained parallel genetic algorithms. Proc. 3rd Int. Conf. Genetic Algorithms, 428–433.
142. Maniatis, T., S. Goodbourn, J.A. Fischer (1987): Regulation of inducible and tissue-specific gene expression. Science **236**, 1237–1245.
143. Maniezzo, V. (1994): Genetic evolution of the topology and weight distribution of neural networks. IEEE Trans. Neural Networks **5(1)**, 39–53.
144. Markel, J.D. and A.H. Gray, Jr. (1976): Linear Prediction of Speech, Springer-Verlag.
145. Martyu L. (1993), MRPII:Integrating the business– A practical guide for managers, Butterworth-Heinemann Ltd (BH Ltd.).
146. MATHWORKS (1991): MATLAB user's guide. The MathWorks, Inc.
147. McFarlane, D.C. and K. Glover (1990): Robust controller design using normalized coprime factor plant descriptions. Lecture Notes Control & Information Sciences, **138**, Berlin:Springer-Verlag.

148. McFarlane, D.C. and K. Glover (1992): A Loop Shaping design procedure using synthesis. IEEE Trans. Auto. Control, **AC-37(6)**, 749–769.
149. Michalewicz, Z. (1996): Genetic Algorithms + Data Structures = Evolution Program. 3rd Ed., Springer-Verlag.
150. Miller, G.F., P.M. Todd, and S.U. Hegde (1989): Designing neural networks using genetic algorithms. Proc. 3rd Int. Conf. Genetic Algorithms, 379–384.
151. Minifie J.R. and R.A. Davis (1990): Internation fffects on MRP nervousness. International Journal of Production Research, **28(1)**, 173–183.
152. Montana, D.J. and L. Davis (1989): Training feedforward neural networks using genetic algorithms. Proc. 11th Joint Conf. on Artificial Intelligence, **IJCAI-11**, 762–767.
153. Mühlenbein, H. (1989): Parallel genetic algorithms, population genetics and combinatorial optimization. Parallelism, Learning, Evolution, Springer-Verlag, 398–406.
154. Munakata, T. and Yashvant Jani (1994): Fuzzy systems: An overview. Communications of the ACM, **37(3)**, 69–76.
155. Munetome, M., Y. Takai and Y. Sato (1993): An efficient migration scheme for subpopulation-based asynchronously parallel genetic algorithms. Proc. 5th Int. Conf. Genetic Algorithms, 649.
156. Myers, C., L.R. Rabiner and A. Rosenberg (1980): Performance tradoffs in dynamic time warping algorithms for isolated word recognition. IEEE Trans. acoustic, speech and signal processing, **ASSP-28(6)**, December.
157. Nambiar, R. and P. Mars, (1993): Adaptive IIR filtering using natural algorithms. Proc. Workshop on Natural Algorithms in Signal Processing, 20/1–20/10.
158. Noll, A.M. (1964): Short-time spectrum and "Cepstrum" techniques for vocal-pitch detection. J. Acoust. Soc. Amer., **36(2)**, 296–302.
159. Noll, A.M. (1967): Cepstrum pitch determination. J. Acoust. Soc. Amer., **41(2)**, 293–309.
160. Nyquist, H. (1928): Certain topics in telegraph transmission theory. Trans AIEE, **47**, February, 617–644.
161. Ogata, K. (1990): Modern Control Engineering. Prentice-Hall International Inc.
162. Omlin, C.W., and C.L. Giles (1993): Pruning recurrent neural networks for improved generalization performance. Tech. Report No. 93-6, Computer Science Department, Rensselaer Polytechnic Institute.
163. Oppenheim, A.V. and R.W. Schafer (1989): Discrete-time signal processing. Prentice-Hall, Englewood Cliffs, New Jersey.
164. Palmer, C.C. and A. Kershenbaum (1995): An approach to a problem in network design using genetic algorithms. Networks, **26**, 151–163.
165. Park, D., A. Kandel and G. Langholz (1994): Genetic-based new fuzzy reasoning models with application to fuzzy control. IEEE Trans Systems, Man and Cybernetics, **24(1)**, 39–47.
166. Park, Y., and H. Kim (1993): Delayed-X algorithm for a long duct system. Proc. Inter-Noise 93, 767–770.
167. Parlos, A.G., B. Fernandez, A.F. Atiya, J. Muthusami and W.K. Tsai (1994): An accelerated learning algorithm for multilayer perceptron networks. IEEE Trans. Neural Networks, **5(3)**, 493–497.
168. S. Pierre and G. Legault (1996): An evolutionary approach for configuring economical packet switched computer networks. Artificial Intelligence in Engineering, **10**, 127–134.
169. Procyk, T.J. and E.H. Mamdani (1979): A linguistic self-orgainizing process controller. Automatica, **15**, 15–30.

170. Rabiner, L.R. and R.W. Schafer (1978): Digital processing of speech signals. Prentice-Hall, New Jersey.
171. Rabiner, L.R., A. Rosenberg and S. Levinson (1978): Considerations in dynamic time warping algorithms for discrete word recognition. IEEE Trans. Acoustic, Speech and Signal Processing, **ASSP-26(6)**, December.
172. Rabiner, L.R. (1989): A tutorial on hidden Markov models and selected applications in speech recognition. Proc. IEEE, **77**, Feburary, 257–285,
173. Rabiner, L.R. (1993). Fundamentals of speech recogntion. Prentice Hall, Englewood Cliffs, New Jersey.
174. Radding, C. (1982): Homologous pairing and strand exchange in genetic recombination. Annual Review of Genetics, **16**, 405–437.
175. T.S. Rappaport (1989): Characterization of UHF multipath radio channels in factory buildings. IEEE Trans. Antennas Propagat., **37**, 1058–1069.
176. I. Rask and C.S. Downes (1995): Genes in medicine. Chapman & Hall.
177. Reed, F.A., P.L. Feintuch, and N.J. Bershad (1981): Time-delay estimation using the LMS adaptive filter-static behavior. IEEE Trans. Acoust., Speech, Signal Processing, **ASSP-29**, 561–568.
178. Reinfeld N.V. (1982): Production and inventory control. Reston Publishing Company, Inc,.
179. Richardson, J.T., M.R. Palmer, G. Liepins and M. Hilliard (1989): Some guidelines for genetic algorithms with penalty functions. Proc. 3rd Int. Conf. Genetic Algorithms, 191–197.
180. Robinson, A.J., J. Holdsworth, R. Patternson and F. Fallside (1990): A comparison of preprocessors for the Cambridge recurrent error propagation network speech recognition system. Proc. of Int. Conf. Spoken Language Processing, Kobe, Japan, November.
181. Roe, D.B. and J.G. Wilpon (1993): Whiter speech recognition: The next 25 years. IEEE Communications Magazine, November, 54–62.
182. Rubio, F.R., M. Berenguel and E.F. Camacho (1995): Fuzzy logic control of a solar power plant. IEEE Trans. on Fuzzy Systems, **3(4)**, 459–468.
183. Rudell, R. and R. Segal (1989): Logic synthesis can help in exploring design choice. 1989 Semicustom Design Guide, CMP Publications, Manhasset, NY.
184. Rumelhart, D.E., G.E. Hinton and R.J. Williams (1986): Learning internal representations by error propagation. Parallel Distributed Processing: Explorations in the Microstructures of Cognition, D.E. Rumelhart and J.L. McLelland, Eds. Cambridge, MA: MIT Press, 318–362.
185. Saenger, W. (1984): Principles of nucleic acid structure. Springer Verlag, New York.
186. Safayeui F., L. Purdy, R. Engelen and S. Pal (1991). Difficulties of just-in-time implementation: A classification scheme. International Journal of Operations and Production Management, **11(1)**, 27–36.
187. Sakoe, H. and S. Chiba (1971): A dynamic programming approach to continuous speech recognition. Proc. of Int. Cong. Acoustic., Budapest, Hungary, paper 20C-13.
188. Sakoe, H. and S. Chiba (1978): Dynamic programming algorithm optimization for spoken word recognition. IEEE Trans. on acoustic, speech and signal processing, **ASSP-26**, Feburary, 43–49.
189. A.A.M. Saleh and R.A. Valenzuela (1987): A statistical model for indoor multipath propagation," IEEE J. Select. Areas Commun., **SAC-5**, Feb, 128–137.
190. Sarker B.R. and J.A. Fitzsimmons (1989): The performance of push and pull systems: A simulation and comparative study. International Journal of Production Research, **27**, 1715–1732.

191. Sarker B.R. and R.D. Harris (1988): The effect of imbalance in a just-in-time production systems: A simulation study. International Journal of Production Research, **26(1)**, 1–18.
192. Schaffer, J.D. (1985): Multiple objective optimization with vector evaluated genetic algorithms. Proc. 1st Int. Conf. Genetic Algorithm, 93–100.
193. R. Sedgewick (1990): Algorithms in C.' Addison-Wesley.
194. S.Y. Seidel, T.S. Rappaport (1992): 914MHz path loss prediction models for indoor wireless communications in multifloored buildings. IEEE Trans. Antennas and Propagation, **40(2)**, 207–217.
195. Shannon, C.E. (1968): A mathematical theory of communication. Bell System Tech. Journal, **27**, 623–656.
196. Sharpe, R.N., M.Y. Chow, S. Briggs and L. Windingland (1994): A Methodology using fuzzy logic to optimize feedforward artificial neural network configurations. IEEE Trans. Systems, Man and Cybernetics, **24(5)**, 760–768.
197. H.D. Sherali, C.M. Pendyala and T.S. Rappaport (1996): Optimal location of transmitters for micro-cellular radio communication system design. IEEE J. Select. Areas Commun., **14(4)**, 662–672.
198. Shynk, J.J. (1989): Adaptive IIR filtering. IEEE ASSP Magazine **April**, 4–21.
199. Sidney J.B. (1977): Optimal single-machine scheduling with earliness and tardiness. Operations Research, **25**, 62–69.
200. Simpson, P.K. (1990): Artificial neural systems: Foundations, paradigms, applications, and implementations. Pergamon Press, 100–135.
201. Simth D. (1985): Bin packing with adaptive search. Proc. Int. Conf. Genetic Algorithms and Their Applications, 202–206.
202. Skogestad, S., M. Morari and J.C. Doyle (1988): Robust control of ill-conditioned plants: High-purity distillation. IEEE Tran. Auto. Control, **AC-33(12)**, 1092–1105.
203. So, H.C., P.C. Ching, and Y.T. Chan (1994): A new algorithm for explicit adaptation of time delay. IEEE Trans Signal Processing, **42(7)**, 1816–1820.
204. Spears, W.M. and K. DeJong (1991): An analysis of Multi-point crossover. Foundations of Genetic Algorithms, G.J.E. Rawlins (Eds), 301–315.
205. Srinivas, M. and L. M. Patnaik (1994): Genetic algorithms: a survey. Computer, June, 17–26.
206. Sugimori Y., K. Kusunoki, F. Cho, K. and Uchikaa (1977): Toyota production system and Kanban system materialization of just-in-time and respect-for-human system. International Journal of Production Research, **15**, 553–564.
207. Sutton, T.J., S.J. Elliott and A.M. McDonald (1994): Active control of road noise insider vehicles. Noise Control Eng. J., **42 (4)**, 137–147.
208. Syswerda, G. (1989): Uniform crossover in genetic algorithms. Proc. 3rd Int. Conf. Genetic Algorithms, 2–9.
209. Syswerda, G. (1991): Schedule optimization using genetic algorithms. Handbook of Genetic Algorithms, 332–349.
210. Szostak, J., T.L. Orr-Weaver, R.J. Rothstein, F.W. Stahl (1983): The double-strand-break repair model for recombination. Cell **33**, 25–35.
211. Szwarc W. (1993): Adjacent orderings in single-machine scheduling with earliness and tardiness penalties. Naval Research Logistics, **49**, 229–243.
212. Szwarc W. and S. K. Mukhopadhyay (1995): Optimal timing scheduling in earliness-tardiness single machine sequencing. Naval Research Logistics, **21**, 1109–1114.
213. Naofumi Takagi, Hiroto Yasuura and Shuzo Yajima, (1985): High-speed VLSI multiplication algorithm with a redundant binary addition tree. IEEE Trans Computers, **C-34(9)**, 789–796.

214. Tamaki, H. and Y. Nichikawa (1992): A paralleled genetic algorithm Based on a neighborhood model and its application to job shop scheduling. Parallel Problem Solving from Nature, 2, 573–582.
215. Tang, K.S., K.F. Man and C.Y. Chan (1994): Fuzzy control of water pressure using genetic algorithm. Proc IFAC Workshop on Safety, Reliability and Applications of Emerging Intelligent Control Technologies, 15–20.
216. Tang, K.S., K.F. Man and S. Kwong (1995): GA approach to time-variant delay estimation. Proc. Int. Conf. on Control and Information, 173–175.
217. Tang, K.S., K.F. Man, C.Y. Chan, S. Kwong, P.J. Fleming (1995): GA approach to multiple objective optimization for active noise control. IFAC Algorithms and Architectures for Real-Time Control AARTC 95, 13–19.
218. Tang, K.S., C.Y. Chan, K.F. Man and S. Kwong (1995): Genetic structure for NN topology and weights optimization. 1st IEE/IEEE Int. Conf. on GAs in Engineering Systems: Innovations and Applications, Sheffield, UK, 250–255.
219. Tang, K.S., K.F. Man, S. Kwong, C.Y. Chan and C.Y. Chu (1996): Application of the genetic algorithm to real-time active noise control. Journal of Real-Time Systems, **13(3)**, 289–302.
220. Tang, K.S., C.Y. Chan, K.F. Man (1996): A simultaneous method for fuzzy membership and rules optimization. IEEE Int. Conf. on Industrial Technology, Shanghai China, 279–283.
221. Tang, K.S., K.F. Man, S. Kwong and Q. He (1996): Genetic algorithms and their applications in signal processing. IEEE Signal Processing Magazine, **13(6)**, 22–37.
222. Tang, Y.C. (1994): Tolkien reference manual. Dept. of Computer Science, The Chinese University of Hong Kong.
223. Tanse, R. (1989): Distributed genetic algorithms. Proc. 3rd. Int. Conf. Genetic Algorithms, 434–439.
224. Thomas, B. (1992): Users guide for GENEsYs. System Analysis Research Group, Dept. of Computer Science, University of Dortmund.
225. Velickho, V.M., and N.G. Zagoruko (1970): Automated recognition of 200 words. Int. J. Man-Machine Stud., **2**, June, 223.
226. Wang D.W. (1995): Earliness/tardiness production planning approaches for manufacturing systems. Computers & Industrial Engineering, **28(3)**, 425–436.
227. Wang D.W. and C.G. Xu (1993): A comparative study on material requirements planning and just-in-time. Acta Automation Sinica, **19**, 363–372.
228. Weinberg, L. (1975): Network analysis and synthesis. R.E. Kreiger, Huntington, NY.
229. Whidborne, J.F., G. Murad and D-W Gu. and I. Postlethwaite (1993): Robust Control of an Unknown Plant. Leicester University Report 93-53.
230. Whidborne, J.F., I. Postlethwaite and D.W. Gu (1994): Robust controller design using loop-shaping and the method of inequalities. IEEE Trans Control System Technology, **2(4)**, 455–461.
231. Whidborne, J.F., D.W. Gu and I. Postlethwaite (1995): Algorithms for solving the method of inequalities - a comparative study. Proc. American Control Conference.
232. White, G.M., R.B. Neely (1976): Speech recognition experiments with linear prediction, bandpass filtering and dynamic time warping. IEEE Trans. Acoust., Speech and Signal Proc., **ASSP-24**, April, 183–188.
233. White, M.S. and S.J. Flockton (1993): A comparative study of natural algorithms for adaptive IIR filtering. Workshop on Natural Algorithms in Signal Processing, 22/1–22/8.

234. Whitley, D. (1989): The GENITOR algorithm and selection pressure: Why rank-based allocation of reproductive trials is best. Proc. 3rd Int. Conf. Genetic Algorithms (J.D. Schaffer, Ed.) 116–121.
235. Whitley, D. (1993): A genetic algorithm tutorial. Technical Report CS-93-103, Department of Computer Science, Colorado State University.
236. Widrow, B. and S.D. Stearns (1984): Adaptive signal processing. Prentice Hall.
237. Widrow, B., D.E. Rumelhart and M.A. Lehr (1994): Neural networks: applications in industry, business and science. Communication of the ACM, **37(3)**, 93–105.
238. Wienke, D., C. Lucasius and G. Kateman (1992): Multicriteria target vector optimization of analytical procedures using a genetic algorithm. Part I. Theory, numerical simulations and application to atomic emission spectroscopy. Analytica Chimica Acta, **265(2)**, 211–225.
239. Wolberg, W.H., and O.L. Mangasarian (1990): Multisurface method of pattern separation for medical diagnosis applied to breast cytology. Proc. of the National Academy of Sciences, 87, 9193–9196.
240. Wright, A.H. (1991): Genetic algorithms for real parameter optimization. Foundations of Genetic Algorithms, J.E. Rawlins (Ed.), Morgan Kaufmann, 205–218.
241. Youm, D.H., N. Ahmed, and G.C. Carter (1982): On using the LMS algorithm for delay estimation. IEEE Trans. Acoust., Speech, Signal Processing, **ASSP-30**, 798–801.
242. Young, S.J., P.C. Wood and W.J. Byrne (1994): Spontaneous speech recognition for the credit card corpus using the HTK Toolkit. IEEE Trans. Speech and Audio Processing, **2(4)**, 615–621.
243. Young S. (1996): A review large-vocabulary continuous-speech recognition. IEEE Signal Processing magazine, Sept., 45–57.
244. Yuan, Z.D. and X. Wang (1994): Determining the node number of neural network models. IFAC Workshop on Safety, Reliability and Applications of Emerging Intelligent Control Technologies, 60–64.
245. Zadeh, L.A. (1973): Outline of a new approach to the analysis complex systems and decision processes. IEEE Trans. Syst., Man, Cybernetics, **SMC3**, 28–44.
246. Zakian, V. and U. Al-Naib (1973): Design of dynamical and control systems by the method of inequalities. Proc. IEE, **120(11)**, 1421–1427.
247. Zhang, J. (1992): Selecting typical instances in instance-based learning. Proc. of the 9th International Machine Learning Conference, 470–479.

Index

Active noise control, 92
Adenine, 1
Amino acid, 2
Anticodon, 2
Architecture
- MIMD, 48
- SIMD, 122
Asynchronous transfer mode, 274

Base pairing theory, 2
Base station, 295

Chromosome, 1, 7
Codon, 2
Constrained LMS, 85
Crick, 5
Crossover
- analogous crossover, 30
- heuristic crossover, 41
- multi-point crossover, 28, 40
- one-point crossover, 10, 14, 40
- reduce-surrogate crossover, 29
- two-point crossover, 29
- uniform crossover, 29, 40
Cytosine, 2

Deceptive function, 16
Defuzzification, 173
- centroid, 174
Deoxyribonucleic acid (DNA), 1
Digital signal processor, 235
Digital signal procssor, 117
Diploid, 55
DNA, 3, 6
Domination, 51
domination, 52
Dynamic time warping, 196

Filter
- Butterworth, 73
- Chebyshev Type 1, 73
- Chebyshev Type 2, 73
- Elliptic, 73
- FIR, 84, 98
- IIR, 73, 118
Fitness, 7
FPGA, 120
Fuzzification, 170
Fuzzy rule, 171
Fuzzy subset, 170

GA processor, 123
Gene, 1, 7
- Exon, 65
- Intron, 65
GENOCOP, 278
Guanine, 1

H-infinity
DOF, 134
- post-compensator, 133
- pre-compensator, 133
Hidden Markov Model, 197, 212
- continuous HMM, 219
- discrete HMM, 213
Hierarchical genetic algorithm, 66
Holliday, 5
Hyperplane, 13
- order of, 13

Implicit parallelism, 13
Inversion, 31

Key process, 257

Manufacturing system
- ETPSP, 254, 256
-- bottleneck, 257
-- discrete form, 256
- JIT, 252
- MPSP, 254
- MRP, 251

– MRPII, 252
Mating restriction, 54
Maximum composition, 173
Mean path loss, 295
Membership function, 170
MENTOR, 281
Mesh network, 282
Migration
– migration interval, 48
– migration rate, 48
– neighbourhood, 48
– ring, 46
– unrestricted, 48
Minimum inferencing, 172
Mutation
– bit mutation, 10, 15, 41
– frameshift, 7
– frameshift mutation
– – deletion, 7
– – insertion, 7
– missence, 7
– neutral, 7
– nonsense, 7
– point mutation, 7
– – transition, 7
– – transversion, 7
– random mutation, 30

Neural network
– bias, 152
– feedforward, 152
– neuron, 152
Niching, 52
Nucleotide, 1

Padé, 136, 141
Parent, 7
Pareto ranking, 53
Pareto-optimal set, 51
Polypeptide, 5
Population, 7
Promoter, 64
Protein, 1

Random immigrant, 55
Regulatory sequence, 64
Replication, 3
Representation
– gray code, 24
– order-based, 24
– real value, 24
– triallelic, 54
Ribonucleic acid (RNA), 2
– mRNA, 5, 65

– tRNA, 5
RLS, 103
RNA, 3

Schema, 13
– defining length of, 15
– order of, 15
Selection
– bias, 27
– efficiency, 27
– random selection, 41
– roulette wheel selection, 9, 41
– spread, 27
– stochastic sampling with partial replacement, 27
– stochastic universal sampling, 27
– tournament selection, 53
Sharing, 54
Solar plant, 189
Speech recognition system
– recognition mode, 196
Speech recognition system , 194
Splicing, 65
Stability triangle, 77
Structural gene, 64
Survival-of-the-fittest, 9

Terminal, 295
Termination, 10
Thymine, 2
Time warping
– DTW, 206
– GTW, 201, 207
– GTW-RSW, 207
– Hybrid-GTW, 207
– hybrid-GTW, 206
Time warping-DTW, 197
Trans-acting factor, 64
Transcription, 3
Translation, 3
Triggered hypermutation, 55

Uracil, 2
Utterance, 193, 196

VHDL, 122

Warping path
– allowable regions, 199
– endpoint constraints, 198
– local continuity, 199
– monotonicity, 198
– slope weighting, 199
water pump system, 181